China's High-Tech Companies

Case Studies of China and Hong Kong
Special Administrative Region (SAR)

China's High-Tech Companies

Case Studies of China and Hong Kong Special Administrative Region (SAR)

Tai Wei Lim
Soka University, Japan

World Scientific

NEW JERSEY · LONDON · SINGAPORE · BEIJING · SHANGHAI · HONG KONG · TAIPEI · CHENNAI · TOKYO

Published by

World Scientific Publishing Co. Pte. Ltd.
5 Toh Tuck Link, Singapore 596224
USA office: 27 Warren Street, Suite 401-402, Hackensack, NJ 07601
UK office: 57 Shelton Street, Covent Garden, London WC2H 9HE

Library of Congress Control Number: 2023035279

British Library Cataloguing-in-Publication Data
A catalogue record for this book is available from the British Library.

CHINA'S HIGH-TECH COMPANIES
Case Studies of China and Hong Kong Special Administrative Region (SAR)

ISBN 978-981-128-097-9 (paperback)
ISBN 978-981-128-098-6 (ebook for institutions)
ISBN 978-981-128-099-3 (ebook for individuals)

For any available supplementary material, please visit
https://www.worldscientific.com/worldscibooks/10.1142/13532#t=suppl

Desk Editors: Logeshwaran Arumugam/Kura Sunaina

Typeset by Stallion Press
Email: enquiries@stallionpress.com

Printed in Singapore

https://doi.org/10.1142/9789811280986_fmatter

About the Author

Tai Wei Lim is a Professor of Business at Soka University, Japan. He teaches courses on international business, Japanese business management, Japanese popular cultural industries amongst others. He is also an area studies specialist in Northeast Asia, and a professional historian specializing in Japanese and Chinese Studies.

About the Contributors

Yoshihisa Godo is a Professor of Economics at Meiji Gakuin University, Tokyo. He received his PhD degree from the University of Kyoto in 1992. His research fields include development economics and agricultural economics. Godo's *Development Economics* (3rd edition), co-authored with Yujiro Hayami and published by the Oxford University Press in 2005, is especially well known. His Japanese book, *Nihon no Shoku to Nou* (Food and Agriculture in Japan), received the 28th Suntory Book Prize in 2006, one of the most prestigious academic book prizes in Japan.

Yee Lam Elim Wong received her PhD in Japanese Studies from the Chinese University of Hong Kong. Her major research interest is on overseas Chinese history in Japan, and she is also interested in East Asian popular culture and history.

Contents

Part I
Introduction

Chapter 1

Introduction

China is now considered a tech superpower in many areas. This writing selects some aspects and case studies of China's technological developments for further analysis in various areas. They include coal energy, housing, connectivity and digital and space technologies. As a further value-add, this volume will also examine technological development in the periphery of China, focusing on its special administrative region (SAR) of Hong Kong. It does not pretend to be comprehensive in its coverage but surveys a spectrum of sectors in China and Hong Kong to get an idea of their developments. By peering into China through the mainland continental perspective and also looking into China from its periphery (e.g. "Greater China" perspectives from Hong Kong), this volume hopes to provide readers with the broad contours and outlines of technological development in China through a multidisciplinary area studies perspective.

Energy and Environmental Technologies: After the introduction chapter, pre-modern and contemporary historical contexts of technological development in China, this volume studies the factors that power China and technological shifts in favour of environmentalist priorities. In terms of its coal energy use, the People's Republic of China (PRC) greatly shapes global coal prices as well as supply and demand through its immense use of the resource to produce coke, crude steel, pig iron and cement. To mitigate pollution, China is following a regional trend of shutting down small-scale coal-powered generators in favour of larger and more efficient power generation plants. Infrastructure evaluations and political campaigns in the Chinese coal industry, the socioeconomic effects and environmental impact of coal use have also become significant topics in political discussions.

When Chinese coal supply was reduced in the fall of 2016 through various means such as the restriction of trucking venues, prices of coal in China went up but, when the restrictions have been lifted to cope with demand, the increase in coal supply caused coal prices to drop. There is hence a need to mitigate price fluctuations. Additionally, China's energy demand relies on its own coal resources and growing imports. Three provinces stand out in domestic coal supply: Shanxi, Shaanxi and Inner Mongolia.

The "Made in China 2025 blueprint" inked by the State Council in May 2015 aims to modernize/upgrade China's manufacturing sector through focusing on coal utilization in big-scale power and heat generation facilities. The "Energy Development Strategy Action Plan (2014–2020)" published by the State Council prescribes restrictions on annual primary energy and coal consumption until 2020. Measures designed to cope with excess overcapacity in steel and coal production have generated images and concerns of a declining coal belt, raising concerns about the local economies of the affected coal mining towns.

Coal however remains responsible for the majority of China's electricity power generation in spite of large hydroelectric dam constructions and initiatives to implement solar panels and wind turbines. Chinese coal investments overseas have increased to procure more resources to generate greater electricity volume. The United Nations lists China as the world's biggest investor in renewable energy projects since at least 2011. North Korea is an important supplier of coking coal for steelmaking in China. Cheap North Korean coal however comes with political repercussions for China from the international community. China's coal use will have global and regional implications, alongside its connectivity infrastructure construction.

With greater confidence and capabilities, the chapter "China's Capacity-Building Roles in ASEAN's People-to-People Connectivity" indicates China's main aid provision activities are based on the central basic principle of enhancing people's mobility: mobility geographically and mobility across socioeconomic classes through education. The intangible aspects include culture and historical/heritage conservation that cannot be easily quantified in some ways. Within the Association of Southeast Asian Nations (ASEAN) Masterplan for Connectivity, tourism is considered as a "sectoral body" contributing to overall integration. Tourism is a functional entity with a very pragmatic purpose of facilitating overall regional integration plans. Besides policy changes to increase connectivity and intra-ASEAN transport agreements, technology can also be harnessed to increase awareness of pan-ASEAN culture and heritage. China's expertise offered through the Belt and Road Initiative (BRI) and Asian Infrastructure Investment Bank (AIIB) will be useful for achieving the tangible aspects of ASEAN aspirations for people-to-people connectivity. Funding, technology and expertise like building high-speed railways (HSRs) can help physically connect regions and disparate areas like far-flung islands.

China can help by relaying its developmental experiences in environmental clean-up to ASEAN countries. Japan as the earliest postwar East Asian economy to experience fast-growth in the 1960s and China which underwent this transition (in the 1980s to the first decade of the 21st century) are both potential references for the other emerging economies to study when it comes to both positive and negative effects of fast growth. Besides the wider East Asian region, China's technologies also contribute to projects within its intra-regions like the SAR of Hong Kong. Here, housing projects with designs generated in Hong Kong and pods constructed in Shenzhen carry the potential for technological collaboration between the two entities. It may provide a solution to the vexing housing shortage problem in Hong Kong. Hong Kong can also tap into mainland Chinese technologies for creating a clean and green environment in terms of infrastructure construction.

In terms of any discussions on the environment and the impact of food production on it, it would not be possible to decouple from analysing China's agricultural technological development. Since 2013, China has been looking at self-sufficiency in staple grains as a political ideology and a geopolitical strategy to break out of political isolation from hostile nations. In the 14th Five-Year Plan (FYP), China is trying to create agricultural growth by allocating more farmland space for cultivation and augmenting the crop harvests through the use of technologies that improve crop and livestock management (within the context of arresting shrinking agricultural land due to urbanization). The "No. 1 document" 2021 has the objectives of bringing about a more self-reliant domestic seed industry, closer integration with manufacturing and service industries to diversify rural incomes of farmers, encourage revitalization of Chinese rural regions, emphasize grain security, augment resources centres for crops/poultry/livestock/marine fishery and provide long-term support to major breeding programmes.

In the post-pandemic era, China hopes to have consistent outputs of soybeans and edible oilseed crops like rapeseed and peanut to hedge against unstable international supplies of (or disruptions in) edible oils, and practice diversification of agricultural product imports while constructing a modern animal farming system and shoring up output capacity of pigs. Agricultural security in Chinese terms does not refer to ostracizing global imports but rather to maintaining a minimum quota of grain supply from domestic sources which it has already achieved while tapping into a "two markets and two resources" system, drawing resources from domestic and overseas sources. Smart farming technologies, such as drones, satellite imaging and pattern modelling, enable agriculturalists to tap into their smartphones and use intelligent environmental tools to guide their agricultural efforts, keep fertilizers/pesticides use to a minimum and conserve water. Agriculture digital services enable product traceability, provide consumers' capabilities to scan QR codes for information on farmland location, harvest seasons/dates and environmental sustainability.

The Chinese authorities are also very interested in implementing land trusteeships. This complements the concept of modern agribusinesses which taps into the economy of scale to churn out greater output production from large farms through the merger and consolidation of smaller farms, while utilizing farmland trusteeship service organizations to take over and farm the lands from farmers who cannot or are unwilling to do so. China does not want to match American farms in size because China's staple crops of corn, rice and wheat all yield the optimal amounts of food per acre at modest-sized farms, and consolidating large numbers of small farms may create social instability by resettling millions of farmers. There is also an emergence of lifestyle-choice urban farmers. Some wealthy urban dwellers have a lifestyle preference leaning away from reliance on industrial farming due to the trust factor.

Industry 4.0 and Emerging Technologies: From heavy industrial technologies like coal and infrastructures, the following two chapters in this volume shift focus to look at technologies of the future: digital technologies and space programmes (an emerging technology detailed in the epilogue). The chapter on digital technologies will focus on the dynamic duo of ZTE and Huawei and their issues and challenges. ZTE and Huawei represent China's best opportunities for global branding, cutting-edge technologies and, maybe even one day, self-reliance based on indigenously developed technologies. To whither down Chinese appetite for US high-tech products, the Chinese digital tech giants also invested lavishly in cutting-edge research and development (R&D). Huawei, for example, has its own R&D division known as HiSilicon Technologies.

Huawei was a latecomer into the smartphone scene but caught up very quickly and, by 2010, it was able to launch its own smartphone in the open market. Its IDEOS smartphone was a success. ZTE's achievements made it a remarkable company. In 2017, at the peak of its industrial development, ZTE attained the global number one position in the World Intellectual Property Organization (WIPO) ranking system for patents applications list. Huawei's achievements were eclipsed by accusations of espionage, particularly from the West. The US disallowed Huawei equipment into the country. AT&T backed out of selling Huawei telecommunication products in early 2018.

In March 2018, ZTE admitted that it had sold technologies to Iran and North Korea despite the United Nations sanctions against such activities. The daughter of the founder of Huawei, Meng Wanzhou, was detained in Canada at the US request for possible violations of sanctions against Iran. Shortly after Meng's detention, the Chinese authorities detained four Canadians (a diplomat on No Pay Leave, a businessman, a teacher and a drug dealer/user), sparking a row between Canada and China. They have all since been released with help from effortful diplomacy and exchange of detainees.

Initially, US alliance networks had some minor differential views when it came to Huawei's equipment. United Kingdom was convinced that Huawei's equipment should be removed from the core British communication system but its peripheral equipment like towers could still be utilized. Germany subsequently took a similar position. After half a decade of the US–China trade war, Huawei has exited the European market and is banned from India and the US. The ability of ZTE or Huawei to stay at number one or even as a leading contender in smartphone technologies will depend on whether they can keep up with the latest trends. The playing field will include the rollout of the latest technologies that include the likes of 5G technologies, foldable phones and Augmented Reality (AR) devices, where the next generation of technologies and major tech players will compete. The final chapter looks at "China's Space Programme: The Final Frontier."

After the first section on mainland China covering case studies of heavy industrial technology (coal), construction technology (connectivity projects and housing), industry 4.0 technologies (digital tech giants) and emerging technologies (space exploration), the subsequent section examines the case study of Hong Kong and their tech companies, policies and developments.

Chapterization

In terms of chapterization, this volume will be examining the development of technologies by the state and private sectors in China and Hong Kong in selected industries and sectors. The first chapter "Introduction" provides some pre-modern historical context in Chinese technological development through a case study in the navigation and maritime sector. The chapter on "Contemporary China's Economic Reforms and Technological Development" brings the story of technological development to the contemporary period. The chapter "Coal Companies, their Technologies and Importance in China's Energy Mix" is useful to examine how China powered its contemporary technological development, starting with the hydrocarbon energy source that it utilized most during its post-1949 industrialization. Chinese coal companies and their extraction technologies formed the backbone of energy provision in China.

The chapter "Technologies and Modernization in China's Farm Development" examines the role of Chinese farms and companies in developing their country's agricultural sector through the use of technologies and farm reforms. As the Chinese people's incomes increase, their consumers' expectations of agricultural product quality also increases. To provide some comparative perspective, newer-age eco-farming and production in China and its neighbour Japan are included in this volume. The chapter titled "Comparative Study of Eco-Feeds: A New Type of Small-Sized Poultry Farming in Japan (Eco-feeding at Yokomine's Farm in Osaka) with Comparative References to China" by Yoshihisa Godo benchmarks the eco-feed in East Asia's most developed agricultural sector in Japan with China's emerging eco-feed industry and uncover how technologies have made China a major player in this field.

Besides serving internal domestic consumers, China is also actively looking towards exporting its high technologies. The chapter "China's Capacity Building Roles in ASEAN's People-to-People Connectivity" looks outwards into another region through China's exportation of technologies to other regions by its state and private sectors. With rapid technological development, China, with its state and private sectors, is now ready to assist other countries with their own developments. Perhaps the region that stands to gain from this is found in its own backyard, the Southeast Asian region nearest to China. This chapter examines such roles in the context of ASEAN's desire for connectivity.

The following section of this volume examines the latest developments in China's SAR of Hong Kong. Hong Kong is probably one of the frontline interfaces that China has in dealing with the global tech industries. It is therefore useful to study the policies of the SAR government. The chapter "The Tech Hub of the Hong Kong Special Administrative Region (SAR)" examines Hong Kong's policy role in luring high-tech companies to situate in the SAR, including those from mainland China. While Hong Kong's tech policies help us understand the local government's orientation towards the tech industries, it is useful to contextualize Hong Kong's positionality within the macro Chinese technological framework, in particular, the Greater Bay Area (GBA) region that Hong Kong is situated in. The chapter "Greater Bay Area Technological Development" Examines Hong Kong's GBA macro-region for its techno-economic future. The nexus between Hong Kong and China's scientific development is examined here.

The two macro chapters on Hong Kong's positionality in the wider context of Chinese development are followed by a case study that parallels the earlier chapter on China's maritime navigation tech development. The chapter on "Hong Kong's Maritime Technologies" details the development of Hong Kong as a global port city with sophisticated container traffic and shipping activities, including the roles played by its shipping companies and public policy. The section on Hong Kong wraps up with the chapter on "Utilization of Advanced Technology During the COVID-19 Pandemic in Hong Kong" which details the technologies that Hong Kong used to mitigate COVID-19.

Finally, this volume examines science's final frontiers in space exploration in its epilogue. The final chapter in this volume looks at the development of science and technology in the arena of space exploration and the progress that the Chinese space programme has made over the years.

Pre-modern Historical Background: Some noted from the world's historical viewpoint that China was the world's leading manufacturer for 1500 years until approximately 1850 when Great Britain surpassed China in the second industrial revolution (the period that roughly corresponded with the invention of steam engines to institutions of mass production).[1] To better understand China's historical development, it may be useful to look at China's pre-modern scientific and technological developmental history, through the lens of a case study. An example of an early technology that China excelled in was maritime navigational technology which can provide a longitudinal historical development of one particular Chinese tech sector. From this case study, it is possible to detect a long history of technological development in China.

Case Study: History of Pre-modern Chinese Maritime Technologies[2]

Chinese primitive technologies in maritime navigation

The Use of Bamboo: In early pre-modern history, Chinese shipbuilders utilized bamboos as flotation materials to build rafts due to its strong longitudinal strength and light hollow interior that can be enhanced with septa for reinforcing transverse strength.[3] The Chinese found that if a bamboo stem is divided into two, the nodal septa can potentially become a watertight bulkhead and utilized as a half-piece to construct a boat, and the bamboo material itself with its tensile strength became experimental material for the Chinese to work on.[4]

Oftentimes, ship designs and maritime technologies became more sophisticated due to the necessity of warfare and so did their enemies. For example, the Song naval forces faced formidable enemies whose naval ships were built with high decks that could navigate up to a riverside citadel, allowing enemy naval officers to easily climb to the top of the city's

---

[1] Hout, Thomas and Pankaj Ghemawat, "China vs the World: Whose Technology Is It?" dated December 2010 in *Harvard Business Review (HBR)* (downloaded on 1 January 2022). Available at https://hbr.org/2010/12/china-vs-the-world-whose-technology-is-it.

[2] Derived from a small section of the author's works on maritime silk road.

[3] Xin, Yuanou, "The Mystery of Chinese Ancient Ship" dated 1998 in Korea Science website (downloaded on 1 January 2020). Available at https://www.koreascience.or.kr/article/JAKO199811921924828.pdf, p. 79.

[4] *Ibid.*

high walls from the decks of their ships to begin their invasion.[5] But the Song warships were no pushovers either. Song battleships were armed with firebomb catapults and incendiary arrows with gunpowder tips, and shielded weapons stations on upper decks for crossbowmen who also played the role of watchmen.[6]

China's maritime technologies are drawn from a variety of sources, including literary works, paintings, official records and other forms of documentation. Very often, Chinese records indicate some textual evidence of pre-modern warships of the past, e.g. those found in war narratives in the iconic Chinese classical literary work "The Water Margin" or in combat manuals like "The Essentials of Military Arts", in which warships were portrayed to carry flags bearing the word "commander" and being steered by a big rudder.

Among the array of technologies, the rudder, watertight compartments in the hold, and mat-and-batten sails were well-known Chinese technological innovations, which were items that were detailed by eminent Chinese scientific/technological historian Joseph Needham in his seminal writing "Science and Civilization in China".[7] Other knowledge about Chinese maritime technologies was simply passed down from generation to generation. Many ports became important locations for the shipbuilding and maintenance industries. For example, Ningde in Fujian province in China was an important harbour and shipbuilding dock in the historic Maritime Silk Road (MSR) and inherited the knowledge/skills legacy of shipbuilders for Admiral Zheng He's treasure ships.[8]

Technologies with Palmiped Emulation: Another observation made by some historians is that the Tang/Song ships had flat bottoms with deadrise and a waterline resembling that of a Palmiped waterbird floating on water and had their widest beam located after amidship.[9] Joseph Needham cited

---

[5] Ebrey, Patricia Buckley, University of Washington, "Warships" undated in the A Visual Sourcebook of Chinese Civilization, University of Washington website (downloaded on 1 January 2020). Available at https://depts.washington.edu/chinaciv/miltech/warship.htm.
[6] *Ibid.*
[7] Xin, Yuanou, *op. cit.*, p. 78.
[8] Ma, Zhou, "Tianjin University Establishes "Database" for Rebuilding Ancient Ships" dated 2016 in Tianjin University website (downloaded on 1 January 2020). Available at http://www.tju.edu.cn/english/info/1012/1186.htm.
[9] Xin, Yuanou, *op. cit.*, p. 80.

Admiral Paris in 1840 as the first observer to note that Chinese shipbuilders imitated their ship designs after living things like Palmipeds that floated on the water surface with the widest breadth of bottom contact with the water and, in doing so, noted that birds operate in the two mediums of air and water while fish only navigated through water.[10]

The Hai Peng junk built by Superintendent of Shanghai shipbuilding Luo Bi (Wade-Giles Lo Pi) in the Yuan Dynasty (1271–1368 AD) and the riverborne junks of Mayangzi, Mayanghuzi, Jinyinding, Hongxiuxie and Yaoshao (in Hubei) all carried the waterplane layout like a waterbird that can fly off a water surface at high velocity.[11] Luo Bi's design philosophy was to have a ship with a "tortoise body and snake head," something adapted from designs originating from dynasties of the past, a design that some scholars argued led to the use of screw propellers in place of paddles.[12]

The Yuloh Paddles: The Chinese had executed all efforts in imitating Nature for implementing stern propulsion rather than relying on the towing force to the prow.[13] The use of robust paddles (the Yulohs) placed at the port/starboard of the stern emulated the webbed feet of a Palmiped waddling in water; the Yulohs were used since the Chinese Western Han Dynasty (206 BC–8 AD) for propulsion purposes.[14]

The Mat-and-Batten Technology: The Chinese invented sails that moved, as opposed to the ancient Mediterranean sails that needed to wait for favourable wind changes to sail. Chinese sails were trimmable so that their ships were able to sail without taking into consideration where the winds were blowing.[15] The Chinese mat-and-batten longitudinal sail had a straight fore edge with inclined hard and a four-cornered balanced stiffened lugsail area forward of the balancing mast, taking up between 1/6th to 1/3th of the overall sail area.[16]

---

[10] *Ibid.*

[11] Xin, Yuanou, *op. cit.*, p. 81.

[12] *Ibid.*

[13] Xin, Yuanou, *op. cit.*

[14] Xin, Yuanou, *op. cit.*, pp. 80–81.

[15] PBS, GBH and WBGH Educational Foundation, "China's Age of Invention" dated 29 February 2000 in the PBS website (downloaded on 1 January 2020). Available at https://www.pbs.org/wgbh/nova/article/song-dynasty/cannoj.

[16] Xin, Yuanou, *op. cit.*, p. 82.

The mat-and-batten system had a complex array of bight sheets, leads/ pulleys for the longitudinal sail to rotate around a vertical mast axis without restrictions which made it convenient to operate, and a fore-and-aft batten sail attached to the mat to mitigate wind conditions.[17] The Chinese mat-and-batten sails keep the sail vertically flat and the larger the batten, the flatter the sail, allowing the ship to experience the strongest propulsion when sailing in the face of the wind.[18]

When the sailor encounters a squall, he can furl the sail so that it drops down in accordance with gravity to the lowest batten by manipulating the halyard lift without jamming it.[19] Because the battens act as stiffeners, cloth, canvas and mats can be used as materials to manufacture the sails as the stress areas are well-spread out using bights, leads, pulleys and battens.[20] Battens double up as ladders for sailors to climb from deck to mast tip, reach any sail section and remain usable even with holes in them; so Huangpu River junks are made up of only 25% of sailcloth, 25% of holed sails (which are adaptable to gentler winds) while the remaining 50% is patched up with other materials.[21]

Multimast Sails: In the Three Kingdom era (220–280 AD), the Wu Kingdom, with its famous shipyards and maritime navigational skills, saw Sun Quan building a powerful navy and despatched emissaries to Southeast Asia (Nanyang) for trade and adapted Southeast Asian-type multimast sails for their own ships.[22] The watertight bulkhead facilitates mast erection in the fore-and-aft direction at any transverse position either at port or starboard and can set as well as manipulate the sails without disruptions to the operations.[23]

When Venetian travellers Marco Polo escorted Princess Kuokuo to Persia in 1292 AD for her wedding ceremony, the Mongol ruler Kubilai Khan dispatched a fleet of 14 four-masted ships for the occasion while the succeeding Ming Dynasty operated a navy led by Admiral Zheng He (Cheng Ho in Wade–Giles) with three to nine-masted ships.[24] Right up to

---

[17] *Ibid.*
[18] *Ibid.*
[19] *Ibid.*
[20] *Ibid.*
[21] Xin, Yuanou, *op. cit.*, p. 83.
[22] Xin, Yuanou, *op. cit.*, p. 84.
[23] *Ibid.*
[24] *Ibid.*

the contemporary era, fishing trawlers known as Seven-fans continue to carry seven masts in Tai Lake (Taihu) and there are still five-masted Sha boats in Chang Jing (Yangtze River).[25] Marco Polo was also surprised at the large quantity of ships on Chinese canals/rivers, more than what Polo was used to back in his hometown cities of Italy, like Venice or Genoa.[26]

Use of Watertight Component: The Chinese designed the ship's hold structure, compartmentalizing it into various watertight compartments so that if one compartment had a leakage, it would not affect the hull and cargo.[27] Pre-modern Chinese ships already featured watertight transverse bulkhead compartments which functioned as dividers for holds to keep them dry even when other proximate bulkheads were damaged and enter them.[28] Watertight bulkheads were found in Eastern Han Dynasty (61–230 AD) ships as well as those of the Three Kingdom period (220–280 AD) although, in terms of material artefacts, the most ancient archaeological sample dated back to the Tang Dynasty (618–907 AD).[29]

Third-century Chinese warships had eight compartments.[30] The Tang-era ships had typically 13 compartments separated by solid bulkheads with plugged drain holes under them and when the plug was taken out, water in the bulkheads would be released and, when re-plugged, the compartment would be watertight.[31] Because of the watertight construction, flooding and sinking were less likely and the concept has been an enduring historical contribution to naval designs and maritime safety precautions.[32]

Having been transmitted inter-generationally for more than 650 years, the "watertight capsule technology" was immortalized by the heritage authorities as a national grade intangible cultural heritage item in 2008 and was shortlisted for UNESCO's "list of intangible cultural heritage urgently needed for protection" in 2010.[33]

---

[25] Ibid.

[26] PBS, GBH and WBGH Educational Foundation, op. cit.

[27] Ibid.

[28] Xin, Yuanou, op. cit.

[29] Ibid.

[30] Ebrey, Patricia Buckley, University of Washington, op. cit.

[31] Xin, Yuanou, op. cit.

[32] Xin, Yuanou, op. cit.

[33] Ma, Zhou, op. cit.

The Rudder: The Chinese invented the sternpost rudder in the Han Dynasty, a technology crucial for navigating a ship.[34] Unbalanced steering rudder found at the ship's bottom and bulkhead construction first emerged in Eastern Han Dynasty (25–220 AD) manned by one or many sailors and evolved into a balanced rudder in Song Dynasty (Wade–Giles: Sung Dynasty 960–1127 AD) before becoming a many-holed rudder in Ming Dynasty (1368–1644 AD).[35]

The Tang and Song dynastic rudders were tackled and can be lifted up when the ship moves into shallower waters to prevent grounding.[36] Some of the rudder technologies were observed in artworks that immortalized the emergence of the stern-post rudder. Valerie Hansen who wrote the publication "The Beijing Qingming Scroll and Its Significance for the Study of Chinese History" made the following observations: "The stern-post rudder [was a] steering device mounted on the outside or rear of the hull. [It] could be lowered or raised according to the depth of the water. This type of rudder made it possible to steer through crowded harbors, narrow channels, and river rapids."[37]

The Screw Propeller: As for screw propellers, the earliest mention of such propellers was found in the book Fang Yan (or Fang Yen in Wade–Giles) authored by Yang Xiong (Yang Xiung in Wade–Giles).[38] The Chinese also innovated the Lu screw propeller system in the Western Han Dynasty (206 BC–8 AD) in the form of a curved lengthy oar with invested cup block under its mid-point resting on a thole pin at the stern, acting like a universal joint with its handle connected to a heavy rope on a fixed ring at deck.[39] The push/pull motion on the handle trigger the submerged part of the Lu propeller to navigate through the water at various angles and, at the end of each stroke on the paddle, the rope exerts a jerk motion to move forward.[40] The Lu also cut down water resistance and reduces efforts for

---

[34] PBS, GBH and WBGH Educational Foundation, *op. cit.*

[35] Xin, Yuanou, *op. cit.*

[36] *Ibid.*

[37] Asia for Educators (AFE), Columbia University, "China in 1000 CE the Most Advanced Society in the World Technological Advances during the Song" dated 2022 in AFE Columbia University Singapore (downloaded on 1 February 2022). Available at http://afe.easia.columbia.edu/songdynasty-module/tech-compass.html.

[38] Xin, Yuanou, *op. cit.*, p. 87.

[39] Xin, Yuanou, *op. cit.*, p. 86.

[40] *Ibid.*

the propulsion resulting in the saying "One Lu, equivalent to three oars." When there is even application of swing, push and pull motions, the ship will navigate in a straight direction and it can move sideways when the motions are adjusted accordingly, resulting in the unified propulsion/ steering functions.[41]

Han Dynasty: Very often, maritime technologies were dual-use technologies that were first developed in naval warfare. Naval warfare on a sizable magnitude was visible from the Han Dynasty (206 BC–220 AD) onwards and the Chinese sailors were pioneers in manipulating/navigating their ships with sails/rudders, incorporating compartmentalization, applying rot-proof paint on the ship's bottom and constructing dry docks.[42] By the time of the Warring States, after accumulating knowledge from a long tradition of shipbuilding (one of the oldest in the world), the square ship was double-bodied and composed of two junks tied side by side, appearing in Chinese maritime waters.[43]

Tang Dynasty: In later Tang dynasty and the Five Dynasties and Ten Kingdoms period, the Chinese junk became the most significant sea-going vessel, passenger carriers and cargo vessels plying the Indian Ocean with large spacious hulls, strong construction, and multiple masts and sails, speedy with seaworthiness qualities.[44] Even today, naval architect, marine artist and maritime historian Chung Chee Kit considers the sailing junk as "seakindly and economical … [it] could adapt to stormy weather, was easy to maintain, and easier to manoeuver compared to Western ships."[45]

Another kind of ship that emerged in this period was the wheeled boat. In the Tang and Song Dynasties (618–1127 AD), Che Chuan boats

---

[41] Xin, Yuanou, *op. cit.*

[42] Cole, Bernard D., "The History of the Twenty-First-Century Chinese Navy" dated 2014 in the Naval War College Review, Vol. 67, No. 3 Summer (downloaded on 1 January 2022). Available at https://digital-commons.usnwc.edu/cgi/viewcontent.cgi?article=1292&context=nwc-review, p. 44.

[43] Ebrey, Patricia Buckley/University of Washington, *op. cit.*

[44] Xin, Yuanou, *op. cit.*

[45] The Head Foundation, Chung Chee Kit, "Ancient Chinese Maritime Technologies: How did the Ancient Chinese Sail the Seas?" dated 7 September 2016 in The Head Foundation website (downloaded on 31 December 2021). Available at https://headfoundation. org/2016/09/07/ancient-chinese-maritime-technologies-how-did-the-ancient-chinese-sail-the-seas/.

had semi-submerged wheel blades on port and starboard sides linked to the foot-actuated crankshaft, evolving from basic manual propulsion to manual paddling and from hand sculling to foot pedalling.[46] In terms of warships, sea hawks emerged in the Tang Dynasty and had floating boards on both sides of the ships for stabilization.[47]

Song Dynasty: The Song Dynasty's main contributions to shipbuilding were faster ship velocity, minimization of foundering, adaptability to maritime conditions, and vessels' steadiness.[48] Some in the scholarly community consider China in 1000 CE as "The Most Advanced Society in the World" and that most technological advances were made during the Song Dynasty (Sung in Wade–Giles).[49] The Song Dynasty Chinese had global shipbuilding expertise, churning out ships with watertight bulkheads, enhanced buoyancy and protected cargo.[50] Stern-mounted or stern-post rudders enhanced the steering function while sounding lines were utilized to measure depth and some Song ships with a crew of a few hundred were propelled by oars and sails.[51]

Warfare was also influenced by naval technological advancements. The Song dynastic naval warships utilized sail and paddle-wheel propulsion, including those powered by labourers on treadmills and the warships were armed with long-range projectile missiles.[52] One of the first examples of a cannon in the world date back to the transition period from the Northern to Southern Song around 1127 and the Song used gunpowder to make fire lance flame throwers and other gunpowder weapons like anti-personnel mines.[53] In the Song (Sung in Wade–Giles) Dynasty, warships called sea hawks usually had four to six boards on each side reinforced with iron hulls and constructed with several decks to steady the ship.[54]

The Southern Song naval forces were far bigger and relatively more capable than its Northern Song counterpart, due to the need to prevent northern nomadic invading armies from crossing the Huai and Yangtze

---

[46] Xin, Yuanou, *op. cit.*, p. 85.
[47] Ebrey, Patricia Buckley/University of Washington, *op. cit.*
[48] *Ibid.*
[49] Asia for Educators (AFE), Columbia University, *op. cit.*
[50] *Ibid.*
[51] *Ibid.*
[52] Cole, Bernard D., *op. cit.*, p. 45.
[53] Ebrey, Patricia Buckley/University of Washington, *op. cit.*
[54] *Ibid.*

Rivers to conquer Chinese Han civilization.[55] In the Song Dynasty, as the Chinese naval forces expanded, port facilities, supply depots, docks, marines battle divisions and coast guard divisions were set up.[56] Some technologies in the Song period may have originated in Tang. For example, some scholars like prominent Chinese historian Patricia Ebrey argued that paddle-wheel boats were late Tang inventions but proliferated in use during the Song Dynasty.[57]

In the era of the Southern Song (Sung in Wade–Giles) Dynasty, large 22–28-wheeled paddle Che Chuan warships with their fast velocities and shallow draft were utilized by historical military leaders like the Tai Lake revolutionary leader Yang Yao to defeat the government's navy.[58] Eventually, the government had to take prisoner the rebel's chief ship designer, Gao Xuan, and then well-known General Yue Fei lured the paddles into heavily weeded shallow waters to entangle their wheels before eliminating the lot.[59] Such designs were carried forth into the Qing Dynasty without installing steam engines, unlike the West's paddle steamships.[60]

Perhaps one of the most important inventions in the Song Dynasty period was the compass. One of China's enduring contributions to maritime technologies was the use of the magnetic compass for maritime navigation which started as early as 1044.[61] The Chinese compass based on the concept of a magnetic needle pointing in the north-south direction became essential to maritime travel and was miniaturized during the Song Dynasty (Sung in Wade–Giles, first known artefact dated 1119) and fastened to a fixed stem (instead of floating in the water) in a portable container with a glass cover for maritime travel.[62]

The mariner's compass was absolutely essential to navigate cargo ships carrying products from one location to another in the vast expanses of the seas and oceans. Historically, the compass was created for divination and spiritual objects in the form of a magnetic spoon in the

---

[55] *Ibid.*

[56] Cole, Bernard D., *op. cit.*

[57] Ebrey, Patricia Buckley/University of Washington, *op. cit.*

[58] Xin, Yuanou, *op. cit.*

[59] *Ibid.*

[60] Xin, Yuanou, *op. cit.*, pp. 84–85.

[61] Cole, Bernard D., *op. cit.*

[62] Asia for Educators (AFE), Columbia University, *op. cit.*

Han Dynasty but, when the Song Dynasty started to trade with maritime Southeast Asia after cutting truncated from the overland Silk Road by other Empires like the Mongols, they utilized the compass for indicating sailing direction.[63]

Ming Dynasty: Some argued that the peak of Chinese pre-modern shipbuilding technologies took place during the Ming Dynasty Emperor Yong Le's reign as he commanded 1350 patrol vessels and 1360 warships which were part of a 3800-strong naval fleet.[64] Admiral Zheng He who was also known as Admiral Cheng Ho in Southeast Asia commanded this fleet. The Southern Song ships utilized watertight capsule technology with partitioned cabins turning into watertight capsules that sealed off their connection with neighbouring compartments and they gave Admiral Zheng He's fleet ships the ability to slice through the sea.[65]

The Ming Dynasty naval vessels were also advanced warships complete with archers, cannons and flame weapons and had logistical capabilities as well managed by 12,000 naval officers.[66] The West were the specialists who innovated cannon technologies by casting bronze cannons that surpassed Chinese versions and eventually, the technology was introduced to the Chinese by the Jesuits in the 16th and 17th centuries and the Ming fought the nomadic Manchus by using cannons cast by hired Jesuit priests.[67] These warships that carried cannons which outgunned and out-tonned any other navies in the world were stationed at guard posts and island maritime naval bases, e.g. the Xinjiang (Wade–Giles: Hsinchiangkhon) at Nanjing (Wade–Giles: Nanking) where 400 freighters also called here.[68]

There are attempts to revive and replicate Ming-era naval technologies in China. Longstanding shipyards of the distant yore are now called upon to recreate the ships that used to ply the seas in Admiral Zheng He's fleet. The purposes for doing so range from heritage preservation to pride of place for traditions. The native Ningde shipbuilders trace their predecessors' genesis/origins/work back to the late Yuan Dynasty and early

---

[63] PBS, GBH and WBGH Educational Foundation, *op. cit.*
[64] Xin, Yuanou, *op. cit.*, p. 77.
[65] Ma, Zhou, *op. cit.*
[66] Cole, Bernard D., *op. cit.*
[67] PBS, GBH and WBGH Educational Foundation, *op. cit.*
[68] Xin, Yuanou, *op. cit.*, pp. 77–78.

Ming Dynasty and some have continued this line of work for 23 generations until today.[69]

There are some observers/critics of the Ming ship technologies in the contemporary period. For example, naval architect, marine artist and maritime historian Chung Chee Kit spoke on this topic. He tried to clarify what he believed were the differences between history and myth, probing if Admiral Zheng He's treasure ships were truly 400 feet in length (and considerably bigger than Christopher Columbus' St. Maria that was 85 feet long) and cited the differences between the ratios in official historical records and more technically feasible contemporary records.[70]

Qing Dynasty: By the Qing Dynasty (1644–1911), the Chinese authorities set up a Shanghai arsenal to construct steam-driven modern warships but constantly faced resistance from the Confucianists and conservatives who held sway China's ideological concerns, one that hampered naval professionalism.[71] Li Hongzhang, a reformist scholarly official/mandarin overhauled the Beiyang Fleet into the most modernized and capable fleet, including two 7500-tonned German battleships, and also purchased or reverse-engineered other warships from the West.[72]

While China remained at the frontlines of developing maritime technologies for centuries, the Chinese maritime industry eventually declined in the 19th century due to reasons offered by naval architect, marine artist and maritime historian Chung Chee Kit, including the appearance of Western steamships that no longer relied on the winds which is a quality that is superior for long-distance trade.[73]

The Rise of China as a Maritime Power[74]: After the establishment of the PRC in 1949, the government focused on the maritime activities in the following areas. They were setting up a marine fishery (a national fishing corporation and fishing communes), nationalized salt beds/coastal salt manufacturing, coastal wetlands reclamation, marine transportation, a deep-sea merchant ship corporation, cargo shipbuilding in Dalian/ Shanghai, expanding ports at Tanggu Tianjin/Zhanjiang Guangdong,

---

[69]Ma, Zhou, *op. cit.*

[70]The Head Foundation/Chung Chee Kit, *op. cit.*

[71]Cole, Bernard D., *op. cit.*, p. 47.

[72]*Ibid.*

[73]The Head Foundation/Chung Chee Kit, *op. cit.*

[74]Adapted from a small section of the author's manuscript on Maritime Silk Road.

Shanghai Fisheries College/Chinese Academy of Sciences' Qingdao Marine Biological Laboratory, and naval construction.[75] In September 1958, China's "Government Declaration Concerning the Territorial Sea" declared that China's territorial sea extended 12 nautical miles from the straight baselines, including coastal islands, Taiwan (claimed by China) and the Penghu/Dongsha/Pratas/Xisha/Paracels/Zhongsha/Macclesfield Bank/Nansha/Spratlys Islands.[76] [In the contemporary period, China constructed Sansha City and gave its local government control over 2 million square km (800,000 square miles) of maritime and overland territories and China's claims in the South China Sea (SCS where Vietnam, the Philippines, Taiwan, Malaysia, and Brunei are the other claimants).[77]]

The History of Chinese Scientific and Technological Power: Zhu Kezheng (Coching Chu, 1890–1974, a famous Chinese meteorologist) set up an Oceanic Section Office within the Science Planning Committee that evolved into the modern State Science and Technology Commission that gave rise to the Ministry of Science and Technology in the PRC in 1956.[78] The Ministry is made up of seven sections in physics, deep seas and oceans, hydrometeorology, chemistry, biology, geology/geomorphology and observation instruments and it was responsible for surveying the Bohai, Yellow, and East China Seas in 1958 and in the SCS in 1959 on behalf of the People's Liberation Army Navy (PLAN).[79] In 1959, on the 10th anniversary of the PRC, the authorities founded Shandong College of Oceanology (now Ocean University of China, Qingdao), Institute of Oceanology (Qingdao) and Chinese Academy of Sciences' SCS Institute of Oceanology in Guangzhou, and Navigation Guarantee Department of the Navy HQ's Qingdao 4th Naval Research Institute in charge of hydrography/oceanography.[80]

---

[75] Takeda, Junichi, "China's Rise as a Maritime Power: Ocean Policy from Mao Zedong to Xi Jinping" dated 23 April 2014 in the Sasakawa Peace Foundation (SPF) (downloaded on 23 April 2014). Available at https://www.spf.org/islandstudies/research/a00011.html.

[76] *Ibid.*

[77] Haver, Zachary, "How China is Leveraging Foreign Technology to Dominate the South China Sea" dated 19 April 2021 in Radio Free Asia (RFA) (downloaded on 19 April 2021). Available at https://www.rfa.org/english/news/special/china-foreign-tech-scs/.

[78] Takeda, Junichi, *op. cit.*

[79] *Ibid.*

[80] *Ibid.*

The State Ocean Administration (SOA) was established on 22 July 1964 under the State Council incorporating PLA General Staff Department navy's oceanology institute, marine survey squadron and Oceanic Section Office of State Science and Technology Commission to the SOA before establishing the Qingdao North Sea Bureau, Ningbo/Shanghai East Sea Bureau and Guangzhou South Sea Bureau in 1965.[81] The SOA coastal stations forecasted tidal, coastal weather and wave conditions while operating a former naval survey vessel fleet, constructed scientific instruments with the First Ministry of Machine-Building "to serve in the construction of national defence and the construction of the national economy on the basis of the duties assigned to it by the State Council (especially for oceanographic surveys)."[82] Institutionally, the SOA governs the ocean policy, along with more than 30 party and state organs/agencies in addition to 11 coastal provincial local governments (Liaoning, Hebei, Shandong, Jiangsu, Zhejiang, Fujian, Guangdong and Hainan, municipalities of Tianjin/Shanghai and Guangxi autonomous region).[83]

The 1970s: During the Cultural Revolution, the propaganda slogan "More ships mean more production" led to a glut of dragnet fishing boats that depleted fish in the sea and China reclaimed land for aquaculture, production of papermaking reeds, planting windbreaks trees but they failed to lead more food projection.[84] The SOA operated the indigenously constructed 3,167-ton survey ship Shijian in 1969, the 13,650-ton Xiangyanghong 05 modified Polish freighter in 1972 which ran four surveys from 1976 to evaluate hydrometeorological conditions in the landing site of a transportation rocket (and a long-range ballistic missile) that China tested in the South Pacific in 1980.[85] To get out of its technological and developmental stagnancy, on 18 December 1978, the Chinese Communist Party (CCP)'s highest-ranking body, the Central Committee, declared at its third plenum to get the economy out of chaotic conditions in the aftermath of The Great Helmsman Mao Zedong's Great Proletariat

---

[81] *Ibid.*
[82] *Ibid.*
[83] *Ibid.*
[84] *Ibid.*
[85] *Ibid.*

Cultural Revolution, where scientists, researchers, intellectuals and capitalists were publicly denounced.[86]

Two marine geological survey squadrons of the geology department (current Ministry of Land and Resources) in Tianjin (now in Shanghai) and Zhanjiang carried out geophysical surveys for oil, deploying the Kantan survey ship in the Bohai/Yellow/East China/SCS (especially the Pearl River Delta coast).[87] Mao Zedong passed away in 1976 allowing the CCP to self-reflect on the impact of communism and the Great Proletariat Cultural Revolution[88] and also the immense damage that the Revolution had done to scientific progress and technological advancement. Two years before his death, the SOA also started going into the realm of environmental protection. From 1974 onwards, the SOA began to detect and investigate pollution in the Yellow/East China/SCS and, in Jiaozhou Bay (Qingdao), 40% of shallow waters designated for aquaculture was polluted and toxic oil and fumes rendered seafood inedible.[89] In December 1978, after 10 years the Great Proletariat Cultural Revolution started by the Great Helmsman Mao Zedong created chaos and stimulated paramount leader Deng Xiaoping to initiate dramatic market-oriented economic reforms to expose China to global investors.[90] Thereafter, one starts to see a commodified, commercialized aspect of maritime industrial development.

1980s: The shipbuilding/maritime sectors were consolidated into the Shipbuilding Industry Authority, the Ninth Industrial Management Bureau, and later the Sixth Ministry of Machine-Building, and finally converted into the China State Shipbuilding Corporation (CSSC) state-owned enterprise (SOE) while the Jiangnan Shipyard (Shanghai) and others were converted into corporations.[91] The major cargo-handling ports were upgraded to manage freight while forging highways, railways and river

---

[86]Zhou, Christina and Bang Xiao, "China's 40 years of economic reform that opened the country up and turned it into a superpower" dated 2 December 2018 in ABC (downloaded on 2 December 2018). Available at https://www.abc.net.au/news/2018-12-01/40-years-of-reform-that-transformed-china-into-a-superpower/10573468.

[87]Takeda, Junichi, *op. cit.*

[88]Zhou, Christina and Bang Xiao, *op. cit.*

[89]Takeda, Junichi, *op. cit.*

[90]Zhou, Christina and Bang Xiao, *op. cit.*

[91]Takeda, Junichi, *op. cit.*

links deep into the interior, where port authorities and ship inspection organs manage foreign trade in 25 ports by 1983.[92] The launch of offshore oil/gas fields for exporting energy fuels initially and then used as domestic energy sources with the initial well dug in Bohai Sea and, in 1982, the State Council regulation approved foreign capital and technologies for developing offshore energy resources with China National Offshore Oil Corporation (CNOOC) as the intermediary/partner for such ventures.

After the Marine Environment Protection Law came into effect in 1982, the SOA oversaw surveying/monitoring of the marine environment, undertook scientific research, and environmentally managed offshore oil development and waste disposal.[93] Sometimes, technologies used for environmental monitoring are dual-use tech. Dual-use foreign technological components were integrated into environmental/ecological monitoring equipment, e.g. in 2017, Sansha city bought US Strix Systems wireless self-organizing network for Paracels Qilian Islets sea turtle monitoring, that is in turn integrated into Paracels Tree Island monitoring platform to feed intelligence back to the Woody Island PLA command centre.[94] Supervision, surveying, marine pollution-monitoring of commercial ships and port operations came under the charge of the ship inspection and port regulation department (now part of the Maritime Safety Administration, Ministry of Transport) and care of fishing vessels and fishing ports came under the fisheries law enforcement and fishing port regulation department (now the Bureau of Fisheries, Ministry of Agriculture).[95]

China improved its surveillance sophisticated through acquisitions of foreign technologies in addition to indigenously developed ones. Even in the 21st century, as much as 25% of purchased foreign tech is integrated into Chinese maritime law enforcement vessels, patrol boats, shipborne boats, smaller craft detachable from larger ships, assault boats and unmanned surface vehicles to patrol Sansha territories.[96] In March 1983, SOA's China Marine Surveillance patrol vessels and aircraft conducted their inaugural patrols from Qingdao, Shanghai, and Guangzhou for environmental protection.[97]

---

[92] *Ibid.*

[93] *Ibid.*

[94] Haver, Zachary, *op. cit.*

[95] Takeda, Junichi, *op. cit.*

[96] Haver, Zachary, *op. cit.*

[97] Takeda, Junichi, *op. cit.*

Martin Chorzempa, research fellow at the Peterson Institute for International Economics, highlighted the importance of experimentation in Chinese development during the economic reform period:

> One of the critical elements of that has been experimentation — so they start with something small, learned from that pilot program, and might have multiple iterations of the same pilot program all throughout the country.... They take the stuff that works and scale it up. And the stuff that doesn't work, they stop doing.... That really sets China apart from other countries which often don't do this kind of evaluation of policies. China has continued to do this, and it's been enormously successful.[98]

The economic and technological pragmatism demonstrated during the reform era is still in place today, especially when it comes to piecing together cutting-edge technologies sourced from all over the world, e.g. technologies used to safeguard its interested in disputed territories. Surveillance technologies that China could not produce were purchased off the open market from the US. The Sansha city authorities bought a US counter-surveillance device for cutting off China's communications from eavesdropping from US intelligence agencies tracking China's facilities construction in the SCS.[99]

The 1990s: In the 1980s, above 88% of China's population subsisted on less than $US2 ($2.70) daily and once the economic reforms were unleashed after 1978, the major slogan was "Let some people get rich first," and after almost 40 years of reforms, the figure was less than 6% in 2017 though an income gap opened up.[100] The setting up of Special Economic Zones (SEZs) in the eastern coastal regions of China in cities such as Shenzhen and Xiamen meant that Chinese and foreign multinationals can trade/invest in capitalistic market-oriented areas, quickly transforming areas like the Pudong district of Shanghai, from agricultural areas to coastal industrial hubs by the 1990s.[101] The emergence of a wealthy as well as middle class drove consumption needs and more resources as well as trade were needed to feed this need. The significance

---

[98] Zhou, Christina and Bang Xiao, *op. cit.*

[99] Haver, Zachary, *op. cit.*

[100] Zhou, Christina and Bang Xiao, *op. cit.*

[101] *Ibid.*

of the seas was officially first mentioned at the National Congress of the CPC's 15th Congress in 1997 when then General Secretary Jiang Zemin's report wrote the following:

> The seas are an important element of the national territory and resources that can be developed on an ongoing basis.[102]

In 1999, the CSSC was divided into two SOEs for augmenting competition: the CSSC to manage operations in Shanghai and Guangzhou and the China Shipbuilding Industry Corporation (CSIC) oversaw activities in Dalian and Wuhan.[103]

In addition to China Ocean Shipping (Group) Company (COSCO) founded in 1961, China Shipping Container Lines (CSCL) was established under the China Shipping Group in 1997 and the state initiated the policies of constructing ships domestically and transporting Chinese goods in them.[104] After 1993, China became a net oil importer, acquiring overseas oil fields in the Mid-East, Africa and South America while securing offshore oil and gas fields in the East China and SCS.[105] In 1993, the Chinese acquired a Ukrainian 21,000-ton icebreaker named Xuelong (before constructing their own icebreakers) and the Xuelong carried out five Arctic observation missions from 1999 to 2012, and in 2004, China set up the Huanghe Station observation base on Norway's Svalbard Island to observe Arctic route opening due to melting polar ice cap and resource explorations.[106]

The Rise of China: Reorganization in the 2000s: In the reform era, Shanghai East Bund of Shanghai was a pioneering economic development zones, attracting foreign investments because of lower trade barriers and eventually, China's entry into the World Trade Organisation (WTO) in 2001 signified its emergence in the global economy and by, 2018, it was the largest trading nation globally.[107] Given China's growing international footprint, its maritime reach needed to be increased. In the 16th Congress

---

[102] Takeda, Junichi, *op. cit.*

[103] *Ibid.*

[104] *Ibid.*

[105] *Ibid.*

[106] *Ibid.*

[107] Zhou, Christina and Bang Xiao, *op. cit.*

in 2002, President Hu Jintao announced "the need for a strategic organ to implement maritime development."[108] In the "Outline of the Plan for National Marine Economic Development" implemented in 2003, China revealed a plan to ascend in stages to become a maritime power (a phrase used for the first time) and this ambition was said to be accomplished at the 18th Congress of the CPC in 2012.[109]

The Law on the Administration of the Use of Sea Areas became effective in 2002 (after tentative implementation by SOA and Ministry of Finance), facilitating state ownership of the seas (surface/body/seabed of interior waters/territorial seas) under the State Council and its maritime administrative department (the SOA).[110] The SOA takes charge of functional division of sea for development, administrative protection, nature protection and the sea-use duration (15 years for aquatic breeding, 20 for shipbreaking, 25 for tourism/entertainment, 30 for salt/mineral production, 40 for public interests and 50 years for construction of ports and shipyard factories).[111]

Since former Chinese President and Secretary General Hu Jintao presented the autumn 2012 political report (Hu's final report as leader) at the 18th National Congress of the Communist Party of China, China had declared its ambitions to become a maritime power:

> We should enhance our capacity for exploiting marine resources, develop the marine economy, protect the marine ecological environment, resolutely safeguard China's maritime rights and interests, and build China into a maritime power.[112]

The National People's Congress (NPC) in March 2013 accepted the State Council's overhaul action of the state organs managing ocean policy in one-stop planning/coordination of oceanic issues (including research on medium/long-term ocean development strategies) by the National Oceanic Commission high-level organ.[113] They also unified maritime law enforcement in four organs (China Marine Surveillance under the SOA, Ministry of Public Security's China Coast Guard, Ministry of Agriculture's

---

[108] Takeda, Junichi, *op. cit.*
[109] *Ibid.*
[110] *Ibid.*
[111] *Ibid.*
[112] *Ibid.*
[113] *Ibid.*

Fisheries Law Enforcement Command, and General Administration of Customs' Anti-smuggling Bureau) under SOA's China Maritime Police Bureau/Ministry of Public Security.[114] The surveillance agencies managed to successfully integrate foreign-produced parts into their technologies. For example, Sansha City's unmanned surface vehicle for tracking/intercepting ships from other SCS claimants like Vietnam and the Philippines contained parts from a number of American firms.[115] Director Liu Cigui of the SOA, the Chinese organ/agency in charge of ocean policy explained what being a maritime power means in Chinese terms:

> Building China into a maritime power is an essential path on the way to the sustained development of the Chinese nation and [achievement of the status of a] global power. A 'maritime power' is a country that has great comprehensive strength in terms of the development, use, protection, management, and control of the seas.[116]

One way to maintain power is to utilize advanced Industry 4.0 technologies effectively. Sansha City also has a 7.5 m L30 autonomous surface vehicle (range: 310 nautical miles, top speed: 40 knots, nicknamed the "Look Out") built by Zhuhai Yunzhou Intelligent Science & Technology Co. Ltd (using some American components) and can fit an automatic weapon or precision missile launcher with 5 km range.[117] The Sansha L30 contains 1,635,000 yuan (US$233,571) worth of crucial components from three US and one Austrian company, including a US Navico automatic identification system (AIS) transponder, Airmar weather monitoring components, two Mercury Marine drives and two Austrian Steyr Motors diesel engines.[118] China National Radio broadcasted in 2018 news that the L30 autonomous surface vehicle executed reconnaissance, precision strikes and protecting islands/reefs and their surrounding waters and the technology is promoted as China's "first maritime weapons platform jointly developed by a local private military industry company and a state-owned military industry research institute."[119]

---

[114] *Ibid.*
[115] Haver, Zachary, *op. cit.*
[116] Takeda, Junichi, *op. cit.*
[117] Haver, Zachary, *op. cit.*
[118] *Ibid.*
[119] *Ibid.*

China's maritime ambitions had also been published. Its white paper *Ocean Development Report* (2012) highlights its strategy and policy with regard to maritime power:

> Ocean policy is a code of behaviour established for the state's strategy, course, development plans, and external relations concerning the seas; it is a basic policy embodying the intentions and interests of the state. It encompasses policies concerning development and use of the seas, including utilization of sea areas, development and protection of sea islands, protection of the marine environment, marine science and technology, marine industry, publicity/education, and human resources development. Marine industry includes such fields as transportation, travel/tourism, fisheries, oil and gas development, and manufacturing of engineering equipment.[120]

China's rise as a manufacturing power facilitates its technological progress through the procurement of foreign technologies. Many multinational corporations have Chinese sales offices/subsidiaries, making it difficult to track US tech supply to China, e.g. Sansha supplier Navico US office headquartered in Norway with offices in China or Airmar whose weather-monitoring device is in Sansha's unmanned surface vehicle said:

> …our products are sold and resold through a distribution chain around the world… Airmar does not always have the ability to track where our products have ultimately ended up.[121]

Meanwhile, China's technologies were constantly getting more sophisticated. In some ways, China successfully integrated foreign technologies into its own facilities. US sources indicated that foreign technological components are found in Chinese infrastructure and facilities in the disputed SCS islands under Chinese administration. The US-based Radio Free Asia (RFA) investigation claimed that Chinese government contracts indicated Sansha City (administrative centre of Paracel/Spratly islands) has purchased foreign items. They include hardware, equipment, software, and materials for maritime law enforcement, information security and land and sea surveillance, from at least 25 different US-based

---

[120] Takeda, Junichi, *op. cit.*
[121] Haver, Zachary, *op. cit.*

firms and other countries (Sweden, Austria, Italy, UK, Japan and Taiwan but with majority of the purchases in the US).[122]

In 2011, China built a semi-submersed oil rig, Haiyang Shiyou (Oceanic Petroleum) 981 (30,670 tons), that can drill 3 km deep set up in the SCS.[123] The Chinese are also making headway in Arctic exploration technologies as well. The 5th Xuelong mission made an Icelandic round trip in summer 2012, via the Sea of Japan/Soya Strait/Okhotsk Sea, northern Kurils' Paramushir Island to the Bering Sea before a coastal route through Russia's EEZ, and the melted Arctic ice cap enabled Xuelong to return by the shortest way through central Arctic Sea via the North Pole and Tsugaru/Tsushima Straits.[124] China has been successful in building its own indigenous capabilities while absorbing foreign components.

Emily Weinstein, a research analyst at Georgetown's Center for Security and Emerging Technology, opined that China acquires foreign technology that "involve everything from M&A (mergers and acquisitions) and investments, to copyright infringement and traditional espionage activities, to grey areas like front organizations and United Front operations like professional associations and overseas scholar returnee organizations."[125] To ensure more indigenous components are made for its future technologies, the Made in China 2025 decade-long industrial development plan, but businesses will be an attempt to lessen China's technological dependence on foreign countries, simultaneously heightening suspicions among foreign governments, especially the West, about its intentions and impact on global supply chains.[126]

Moving Forward, Some Concluding Remarks: Ocean-related economic activities made up approximately 10% of China's gross domestic product (GDP) and more than 16% in coastal regions, supplying jobs to 33.5 million.[127] To expand global trade and extraction of world resources, China needed to expand its connectivity into the world region, both overland and

---

[122] *Ibid.*

[123] Takeda, Junichi, *op. cit.*

[124] *Ibid.*

[125] *Ibid.*

[126] tps://www.spf.org/islandstudies/research/a00011.html.
Zhou, Christina and Bang Xiao, *op. cit.*

[127] Takeda, Junichi, *op. cit.*

maritime. 2014–2015 initiated China's road towards the ambitious dollar BRI extending China's political and economic soft power globally internationally as the world witnessed the rise of tech giants like Alibaba and Huawei.[128]

In the 21st century, the SOA printed yearly reports on the marine environment/disasters and they differentiate natural disasters due to such causes as violent winds, abnormally high tides, and icebergs as opposed to environmental disasters like red and green tides (caused by algal blooms), large-scale oil spills, and seawater corrosion.[129] To avoid depletion of fishery resources, fishing permits, fishing boat certs, crew members certification, aquaculture licenses, fishing prohibition areas/seasons, net size regulations, quotas on young fish catches, dynamite/poison/electric shocks bans and single-sweep mass catches were put in place, along with a revised Fisheries Law 2004's quota on catches and anti-corruption measures for fishery officers.[130]

The Sansha headquarters on Woody Island in the Paracel Islands operate China's outposts to implement long-term initiatives with the People's Liberation Army (PLA) in terms of infrastructure, defence capabilities, transportation and communication using a "military-civil fusion" strategy of resource mobilization.[131] Under a one-party system, the Chinese state mobilized resources for non-commercial basic/advanced research like space/military development, national prestige projects and, in maritime research, the indigenous Jiaolong submersible plunged 7062 m into the Pacific Ocean Mariana Trench in 2012 and, in 2011, China launched the Haiyang 2 (HY-2) observation satellite for real-time monitoring of dynamic ocean conditions.[132] China had emerged as an all-round technological power. Jane Golley, acting director at Australian Centre on China in the World at the Australian National University (ANU) opined:

> China's high-tech industries, artificial intelligence, genetic engineering, robotics, space and aviation … are becoming globally competitive and highly innovative enterprises that you just couldn't conceived of two decades ago — obviously not four decades ago. China is far from alone

---

[128] Zhou, Christina and Bang Xiao, *op. cit.*

[129] Takeda, Junichi, *op. cit.*

[130] *Ibid.*

[131] Haver, Zachary, *op. cit.*

[132] Takeda, Junichi, *op. cit.*

on inequality, but it is dire when you've got [so many] billionaires and you're talking robotics, artificial intelligence and world's largest companies and then you've got 84 million people living on under $US2 a day.[133]

But, because of China's expanding footprints in the seas and oceans, it has come up against the world's de facto and current maritime power which is the United States of America (USA). The Chinese state's purchases of foreign-originated technologies for its strategic objectives has resulted in the US government implementing export controls and other means to reduce the flow of technologies to China.[134] Ashley Feng, a specialist on US–China economic relations explained: "Through the Export Administration Regulations, the Munitions List, the Commerce Control List, and the Entity List, the US government can control what technology is exported out of the United States both by where the export will end up and/or whose hands it will end up in."[135] In December 2020, the US Department of Commerce cut off exports to Huazhong Photoelectric Technology Research Institute (also known as CSSC 717th Research Institute and a developer of the L30 autonomous vehicle deployed in Paracel/Spratlys) for "acquiring and attempting to acquire US-origin items in support of programs for the People's Liberation Army."[136]

The West introduced steamship technologies followed by oil-powered ships to China and it facilitated global trade that eventually opened up late pre-modern China into the modern world. Today, China is often said to be the new emerging tech powerhouse that is the peer competitor to the US. China has made public its aspirations to lead in the emerging technologies in the world, including but not only restricted to the all-important field of digital technologies. To understand how contemporary Chinese companies transformed the country into a tech superpower, some background information may be useful. The following chapter will look at how contemporary China develops its technologies.

---

[133] Zhou, Christina and Bang Xiao, *op. cit.*
[134] Haver, Zachary, *op. cit.*
[135] *Ibid.*
[136] *Ibid.*

Bibliography

Asia for Educators (AFE) and Columbia University, "China in 1000 CE The Most Advanced Society in the World Technological Advances during the Song" dated 2022 in AFE Columbia University Singapore (downloaded on 1 February 2022). Available at http://afe.easia.columbia.edu/songdynasty-module/tech-compass.html.

Cole, Bernard D., "The History of the Twenty-First-Century Chinese Navy" dated 2014 in the *Naval War College Review*, Vol. 67, No. 3 Summer (downloaded on 1 January 2022). Available at https://digital-commons.usnwc.edu/cgi/viewcontent.cgi?article=1292&context=nwc-review.

Ebrey, Patricia Buckley and University of Washington, "Warships" undated in the A Visual Sourcebook of Chinese Civilization/University of Washington website (downloaded on 1 January 2020). Available at https://depts.washington.edu/chinaciv/miltech/warship.htm.

Haver, Zachary, "How China is Leveraging Foreign Technology to Dominate the South China Sea" dated 19 April 2021 in Radio Free Asia (RFA) (downloaded on 19 April 2021). Available at https://www.rfa.org/english/news/special/china-foreign-tech-scs/.

Ma, Zhou, "Tianjin University Establishes "Database" for Rebuilding Ancient Ships" dated 2016 in Tianjin University website (downloaded on 1 January 2020). Available at http://www.tju.edu.cn/english/info/1012/1186.htm.

PBS, GBH and WBGH Educational Foundation, "China's Age of Invention" dated 29 February 2000 in the PBS website (downloaded on 1 January 2020). Available at https://www.pbs.org/wgbh/nova/article/song-dynasty/.

Takeda, Junichi, "China's Rise as a Maritime Power: Ocean Policy from Mao Zedong to Xi Jinping" dated 23 April 2014 in The Sasakawa Peace Foundation (SPF) (downloaded on 23 April 2014). Available at https://www.spf.org/islandstudies/research/a00011.html. [Adapted from: Takeda, Junichi, "China's Rise as a Maritime Power: Ocean Policy from Mao Zedong to Xi Jinping" dated 23 April 2014 in Review of Island Studies (downloaded on 1 January 2020). Available at http://islandstudies.oprf-info.org/research/a00011/. Translated from "Chugoku no kaiyo seisaku" dated April 2013 in *Tosho Kenkyu Journal*, Vol. 2, No. 2 (OPRF Center for Island Studies), 2013, pp. 73–95.]

The Head Foundation/Chung Chee Kit, "Ancient Chinese Maritime Technologies: How did the Ancient Chinese Sail the Seas?" dated 7 September 2016 in The Head Foundation website (downloaded on 31 December 2021). Available at https://headfoundation.org/2016/09/07/ancient-chinese-maritime-technologies-how-did-the-ancient-chinese-sail-the-seas/.

Xin, Yuanou, "The Mystery of Chinese Ancient Ship" dated 1998 in Korea Science website (downloaded on 1 January 2020). Available at https://www.koreascience.or.kr/article/JAKO199811921924828.pdf.

Zhou, Christina and Bang Xiao, "China's 40 years of economic reform that opened the country up and turned it into a superpower" dated 2 December 2018 in ABC (downloaded on 2 December 2018). Available at https://www.abc.net.au/news/2018-12-01/40-years-of-reform-that-transformed-china-into-a-superpower/10573468.

Chapter 2

Contemporary China's Economic Reforms and Technological Development

Historical Background

Public ownership of Chinese companies in 1965 made up 90.1% of total industrial production, virtually attaining state ownership of all industrial enterprises, a moment described by the Great Helmsman Mao Zedong in his *The Notes of Political Economy* as the "completeness" of socialism.[1] In the Mao-era Stalinist production superstructure, industrial output and economic activity were disrupted, individual enterprises were banned and converted into collective enterprises and so, with the start of economic reforms in 1978, SOEs accounted for 77.03% of industrial production before economic reforms pared it down to 50% by 1992–1993, by 1996 and 1997, below 30% and 25.52%, respectively.[2] Just before the reform era in 1975, collective enterprises (manufacturing cooperative-owned enterprises in the rural areas and cities-based collective ward-owned enterprises) made up 18.90% of gross industrial production, and after the

---

[1] Kobayashi, Shigeo, Jia Baobo and Junya Sano, "The "Three Reforms" in China: Progress and Outlook" dated September 1999 in Japan Research Institute (JRI) (downloaded on 1 January 2021). Available at https://www.jri.co.jp/english/periodical/rim/1999/RIMe199904 threereforms/.

[2] *Ibid.*

economic and open-door reforms, collective enterprises hit 30% of gross industrial production.[3]

The economic reform era in China implemented a new economic development strategy at the Third Plenary Session of the 11th Chinese Communist Party Central Committee (CCPCC) in late 1978 under the leadership of paramount Deng Xiaoping who returned to power after three political setbacks.[4] The Chinese leadership instituted an open-door policy to encourage the adoption of foreign capital and technology for economic growth, grafting it to a hybrid system that is based on political socialism to construct an economy destroyed by the Great Proletariat Cultural Revolution.[5] Economic reforms and open door economy were adopted to increase the salaries of the average Chinese citizens that were comparatively low relative to other East Asian capitalistic economies and to shore up legitimacy for the Chinese government and the CCP by positioning itself as the legitimate defender of economic growth and betterment of living standards.[6]

The state set up SEZs, economic and technology development/high-tech industry development/delta open/peninsula open zones and open coastal/open border cities zones as experimental test-beds for foreign investments from Hong Kong and Taiwanese firms while it strengthened its own socialist hybrid market economy and this augmented entrepreneurial energies and venture businesses in the country.[7] Foreign funding, technology and know-how transformed its large workforce and developmental space for accelerated economic growth and increased GDP per capita while the open-door economic policy introduced rapid economic expansion in the first five years of the 1980s before an interregnum and continued in the first five years of the 1990s.[8]

The 1990s: With the 1990s transition from dominant state ownership to collective and other ownership structures (individual/private enterprises, foreign-owned/joint-stock companies), by, 1997, there were 98,600 (1.25% of total firms) SOEs out of 7.9 million industrial firms while

---

[3] *Ibid.*
[4] *Ibid.*
[5] *Ibid.*
[6] *Ibid.*
[7] *Ibid.*
[8] *Ibid.*

SOEs made up 25.52% of gross industrial production, 63.52% of net fixed assets of industrial enterprises and 65.0% of employees.[9] A priority in the 1990s was to end textile sector losses that went from 1.9 billion renminbi in 1993 to 10.6 billion in 1996 and over 9 billion in 1997 and by 1996, 42% of state-owned textile firms were losing money, 5 points higher than the ratio of loss-making industrial SOEs in all state-owned industrial enterprises (37%).[10]

Unprofitable textile firms made up 50% of total employees in the state-owned textile sector and a 1991 directive for improving this situation focused on disposing of old textile machines, getting rid of 10 million obsolete textile equipment and targets set for each year (e.g. mandatory phase-out of 4.8 million textile machines in 1998).[11] Progress was slow between 1992 and 1996 with only 210,000 machines phased out because disposed textile machines in the eastern coastal region were relocated to factories in the western inland provinces, thereafter the authorities decided to rely on compulsory disposal and restructuring of textile machines, especially in textile hubs like Shanghai, Qingdao, Tianjin, Wuhan, Jinan and Dailian.[12]

The authorities monitored the sales of domestically retailed textiles through production and purchase permit systems for textile machinery with strict limits placed on textile processing capacity and textile machinery exports were encouraged through the provision of export credit and full refunds of value-added taxes.[13] In this way, the industry became a test base and starting point for SOE reforms. In 1998, the textile industry saw 5.12 million textile machines removed (more than the 4.8 million targets), while the NPC approved *Explanation on the Plan for the Restructuring of the State Council* [into 4 sectors] in March 1998: (1) macro-, (2) industrial/economic control, (3) education/science & technology/culture/social security/resource control and (4) political affairs.[14] In the industrial and economic control sectors, six ministries involved in the control and management of individual enterprises/specific industries were fused into the State Economic and Trade Commission, including the Ministry of Machine-Building Industry and the Ministry of Chemical.[15]

---

[9] *Ibid.*
[10] *Ibid.*
[11] *Ibid.*
[12] *Ibid.*
[13] *Ibid.*
[14] *Ibid.*
[15] *Ibid.*

Then paramount leader Deng Xiaoping's objective of increasing the economic worth of the Chinese people resulted in a 14-times increase from the 1980 figure for per capita income to US\$770 in 1998.[16] But, it also introduced income gaps between the wealthy and the impoverished which contrasted with the comparatively uniform income between different social classes under pre-economic reform communist rule, a reflection of Deng Xiaoping's mantra that it was okay for some coastal provinces to be wealthier than others.[17] Coastal and interior provinces, cities and rural countryside showed the differences between urban incomes and peasant households in backward areas, e.g. Guangdong province income was approximately eight times more than that in Gansu.[18]

SOEs and state-owned banks did not possess globally competitive world-class technology like those of foreign-owned and private enterprises, something crucial given that SOEs made up 41.5% of China's gross industrial production in 1952 while individually owned firms made up 20.6% of production, and companies under other types of ownership accounted for 34.7%.[19] Initially, SOE output in the immediate post-war era originated from companies/industries set up by defeated Axis powers (Imperial Japan, Nazi Germany and Italy) in World War II and were initially nationalized by the Nationalist Kuomintang regime and then appropriated by the CCP after the Chinese civil war.[20]

A Major Technological Superpower: When China hit the 70th year of CCP control of the country, its maturing economy stayed the fastest-expanding globally, continuing to pour funding into technological innovation, with the World Bank (WB) stating: [China] has "experienced the fastest sustained expansion by a major economy in history."[21] As the Chinese economy matures, it is entering a phase of "new normal" of sustainable growth and also looking inwards for domestic consumption as its major economic driver. China is also going to strengthen its ability to

---

[16] *Ibid.*

[17] *Ibid.*

[18] *Ibid.*

[19] *Ibid.*

[20] *Ibid.*

[21] Charlton, Emma, "6 things to know about China's historic rise" dated 1 October 2019 in World Economic Forum (Weforum) (downloaded on 1 October 2019). Available at https://www.weforum.org/agenda/2019/10/china-economy-anniversary/.

innovate and come up with new indigenous technologies that will then be exported overseas, setting global standards for those technologies. Chinese R&D spending increased by 70% higher in 2017 than in 2012, sinking in investments in industrial science parks and incubators working on Industry 4.0 technologies like artificial intelligence (AI), robotics and big data while private sector tech giants are pouring billions in into new research institutes, experts and scientists.[22]

By February 2018, China was behind the US in a number of unicorns (start-ups valued at US$1 billion or higher) and plans to bring the bar even higher through a decade-old Made-in-China 2025 that will accelerate indigenous high-tech development and make the country a global leader in telecommunication, electrical power equipment, robotics, high-end automation and new energy vehicles.[23] Acute state funding and foreign investments have pushed China to double its economic size every 8 years since 1979 through the coastal provinces have unevenly outperformed inland provinces with manufacturing powerhouses like Guangzhou and Shenzhen in Guangdong province the largest GDP contributor to the economy, followed by Jiangsu province.[24]

The stereotypical view of China as a global production site for low-cost products is being deconstructed as the Chinese authorities are trying to augment China's technological ability to produce high-tech products for the world. Some even think that China is on its way to challenge, supplant or overtake the US as the number technological superpower in the world. China also has ambitions to be a global innovation hub. Up till 9 September 2022, there were more than 700 specialized tech firms in the US (market valuation: HK$85.4 trillion) and approximately 450 in China (excluding HKSAR, market valuation: HK$5.3 trillion) with Hong Kong behind numerically by only 100 companies.[25]

The Chinese technology industry enjoys resources provided by the Chinese state. One of the reasons why China's digital technological sector has developed quickly over 20 years in services, communications and

---

[22] *Ibid.*

[23] *Ibid.*

[24] *Ibid.*

[25] Zhang, Tianyuan, "HK to facilitate high-tech floats" dated 21 October 2022 in China Daily (downloaded on 21 October 2022). Available at https://www.chinadaily.com.cn/a/202210/21/WS6351f6c0a310fd2b29e7dbe9.html.

e-commerce is due to the state's urging of Chinese people to utilize digital technologies in their everyday lives and this resulted in the accumulation of a billion Internet users.[26] In another example, the authorities provide sizable tax rebates to local companies like Kingdee International Software Group, shaping it into the largest software supplier to small and medium enterprises (SMEs) in China in the software sector.[27] Between 2000 and 2010, China also increased its R&D expenditures by approximately 21% annually.[28] There are 16 State Key Laboratories (SKLs) under the Chinese Ministry of Science and Technology and six branches of Chinese National Engineering Research Centres in Hong Kong (operated by its public universities/research institutions) while Guangzhou Institutes of Biomedicine and Health (under Chinese Academy of Sciences) established a stem cell and regenerative medicine research facility at Hong Kong's Science Park.[29] In 2006, frustrated by its inability to capture some tech component market shares in its own domestic market and pay higher royalties for those components, China embarked on a Made-in-China programme.[30]

These efforts are not restricted to the Chinese mainland alone. Hong Kong's innovation and technology hub in combination with the Shenzhen and Guangzhou hubs make up the Shenzhen-Hong Kong-Guangzhou science and tech cluster that is globally the second best-performing according to the Global Innovation Index 2021.[31] Hong Kong and Shenzhen signed the January 2017 "Memorandum of Understanding on Jointly Developing the Lok Ma Chau Loop by Hong Kong and Shenzhen" to develop the 87-hectare Hong Kong-Shenzhen Innovation and Technology

---

[26] Gray, Joanne and Yi Wang, "China's big tech problem: even in a state-managed economy, digital companies grow too powerful" dated 13 July 2022 in The Conversation (downloaded on 13 July 2022). Available at https://theconversation.com/chinas-big-tech-problem-even-in-a-state-managed-economy-digital-companies-grow-too-powerful-186722.

[27] Hout, Thomas and Pankaj Ghemawat, "China vs the World: Whose Technology Is It?" dated December 2010 in Harvard Business Review (HBR) (downloaded on 1 January 2022). Available at https://hbr.org/2010/12/china-vs-the-world-whose-technology-is-it.

[28] Ibid.

[29] Fung, Doris, "Innovation and Technology Industry in Hong Kong" dated 6 September 2022 in HKTDC Research (downloaded on 6 September 2022). Available at https://research.hktdc.com/en/article/MzEzOTIwMDIy.

[30] Hout, Thomas and Pankaj Ghemawat, op. cit.

[31] Fung, Doris, op. cit.

Park for luring private sector, research and tertiary institutions.[32] The first phase (2024–2027) will be situated in a 1.2 million square metres facility, the largest HK I&T platform.[33] Hong Kong is also revving up its tech-financing power. Hong Kong's bourse initiative pertains to companies from five "Specialist Technology Industries" of next-gen information technology (IT), advanced hardware/materials, new energy devices, environmental protection infrastructure, new food/agricultural technologies, AI, cloud computing and semiconductors.[34]

Smart City Development: Biotechnology, AI, smart city and financial technologies are major areas in Hong Kong's innovation and tech industry facilitated by an expanding domestic research cluster, a strategic business platform and tech marketplace and the presence of the Guangdong-Hong Kong-Macao GBA.[35] The Hong Kong authorities published Blueprint 2.0 in December 2020[36] containing more than 130 smart city initiatives (e.g. Building Information Modelling, smart tourism facilities, Traffic Data Analytics System, Smart Village pilot, telehealth, video-conferencing/remote consultation, LeaveHomeSafe mobile app).[37] Foreign institutions are also found in Hong Kong. Densely populated Hong Kong is a testing ground for many smart city apps and foreign collaborations like HKSTP-Siemens Smart City Digital Hub at the Science Park for smart city applications specialized in the Hong Kong market.[38]

In Hong Kong, some ranking AI technology private sector firms like SenseTime established in 2014 by Professor Sean Tang and his team are producing a facial recognition system with deep learning exhibiting an accuracy rate of more than 99%, becoming a unicorn within three years (another example is WeLab online lending AI-enabled platform utilizing offering individualized lending packages for customers).[39] In May 2018, Alibaba, SenseTime and HKSTP established the NPO HKAI Lab in Hong Kong with its aspirations to be an international location for AI, push AI boundaries, assist start-ups to commercialize their tech and form

---

[32] *Ibid.*

[33] *Ibid.*

[34] Zhang, Tianyuan, *op. cit.*

[35] Fung, Doris, *op. cit.*

[36] Hong Kong Smart City Blueprint 1.0 was first published in December 2017.

[37] Fung, Doris, *op. cit.*

[38] *Ibid.*

[39] *Ibid.*

ideas/knowledge exchanges.[40] Hong Kong is discussed in greater detail at the end of this chapter.

Autonomous A.I. Ports in China[41]

In addition to smart city development, China is also developing smart ports. Seven of the globe's busiest container terminals are located in China and it is also the first one in the world to build a completely automated and workerless Qingdao New Qianwan Automatic Container Terminal (nicknamed by locals as "Ghost Port") which is located in the eastern Chinese port of Qingdao and became operational in May 2017.[42] The Chinese Ministry of Transportation came up with a blueprint in September 2021 to put in place priority infrastructure like intelligent and unmanned ports. The infrastructure construction was boosted by demand for efficient port solutions and upgrading activities at a time of the COVID-19 pandemic (taking advantage of the pandemic to augment China's smart port manufacturing ability and further boost its share in the global market).[43]

China is advancing with its intelligent augmentation of Xiamen, Ningbo and Dalian container ports while constructing automatic ports in Tianjin, Suzhou and Beibuwan, southwest China's Guangxi Zhuang Autonomous Region and accelerating intelligent scheduling, equipment remote control and unmanned trailer transportation.[44] The Chinese port authorities will also construct smart logistics service platforms and enhanced monitoring technologies for dangerous cargo items and port operators have carried out digitalization of port systems to cut down on the time needed for customs clearance from 1.5 days to just five minutes with the average time taken for container pick-up by 20%.[45] And even

---

[40] *Ibid.*

[41] Derived from a small section of the author's maritime silk road writing.

[42] Han, Peng and Gao Yun, "'Ghost port': Asia's first fully-automated port begins operations in Qingdao" dated 13 May 2017 in CGTN (downloaded on 13 May 2017). Available at https://news.cgtn.com/news/3d637a4e31677a4d/share_p.html.

[43] GT staff reporters, "China to accelerate building of intelligent, unmanned ports in next five years" dated 24 September 2021 in Global Times (downloaded on 24 September 2021). Available at https://www.globaltimes.cn/page/202109/1235005.shtml?id=11.

[44] *Ibid.*

[45] *Ibid.*

with COVID-19 disinfection process, the smart port systems have stronger operational efficiency for moving cargo goods from ships to containers.[46] Moreover, the COVID-19 coronavirus pandemic has resulted in port congestion and less operational efficiency, motivating more local port authorities to upgrade their systems, e.g. Qingdao Port in eastern China Shandong Province declared it was the first in China to attain achieved automatic operation of ship off-loaders in 2021 and it took only three years to construct (reducing investment risks/outlays).[47]

When state-owned media China Global Television Network (CGTN) went to the terminal at midnight, there were no human workers. However, the port equipment and vehicles were operating in the dark off-loading cargo 24 hours ceaselessly from a large-sized cargo vessel, guided AI with the use of laser scanning and positioning to detect the containers' four corners and then place them precisely onto autonomous autopilot vehicles.[48] The smart autopilot trucks powered by electricity have their movements and assignments allocated digitally (including recharging routines), replacing manpower and reducing high labour costs by 70% and logistical bottlenecks at the port's entry point at the same time while operating nocturnally (increasing efficiency by 30%).[49] Zhang Liangang, general manager at the Qingdao New Qianwan Container Terminal, revealed that it needed 60 workers to off-load a cargo ship at a conventional port but the automatic port needed only nine workers, replacing sky cranes human operators with robots.[50]

There is great export potential for Chinese equipment to other countries. In mid-September 2020, US Baltimore Port took delivery of four Neo-Panamax container cranes (137 m tall and 1740 tons each with outreach capacity to manage 85 tons of containers) constructed by Shanghai Zhenhua Heavy Industries Co. (ZPMC, the globe's biggest maker of port equipment) as a component of a blueprint to double the port's handling capacity.[51]

Chinese Tech Sector and its Relationship with the Foreign MNCs: The Chinese authorities have also worked on attracting other companies in the

---

[46] *Ibid.*

[47] *Ibid.*

[48] Han, Peng and Gao Yun, *op. cit.*

[49] *Ibid.*

[50] *Ibid.*

[51] GT staff reporters, *op. cit.*

world to relocate to China. French nuclear power reactors makers, American long-distance planes and Applied Materials (leading supplier of semiconductor-making equipment) parcelled out substantial R&D activities to China and shifted their chief technology officers to China in around 2010.[52] Foreign firms and multinationals made up a majority of China's advanced tech industries, making up 85% of the tech exports from China in 2008 and exports of mobiles and notebooks with less than 10% domestically made components and foreign industries made most of the remaining parts.[53]

In that same period, China has been persuading understatedly and purposefully transitioning shifting its economy from a lower value-added production system to an advanced production landscape by jointly co-opting and sometimes compelling G7 companies to set up shop there.[54] In late 2009, China's Ministry of Science and Technology promulgated that all the technological products retailed to government agencies be developed domestically (it was revoked later after opposition from the foreign private sector) and this may provoke the relocation of MNC R&D activities to China though intellectual property rights (IPR) are not yet aligned with the West.[55]

China's MNC attraction drives were also implemented in its Special Administration Region (SAR) of Hong Kong (HK). Hong Kong's global innovation and technology hub roadmap was also recognized in the 14th FYP (2021–2025) and has positioned the Chief Executive (CE) of Hong Kong Special Administrative Region (HKSAR) to coordinate closely with China. This may involve training new R&D talents locally in Hong Kong. The Research Talent Hub has financed more than 3700 R&D positions, the Re-industrialisation and Technology Training Programme with on-the-job training for 3500 employees from 1800 companies and the GBA Youth Employment Scheme 2021 offering 700 I&T places for Hong Kong's tertiary graduates in I&T in any GBA cities.[56] The Hong Kong authorities instituted the Technology Talent Admission Scheme

---

[52] Hout, Thomas and Pankaj Ghemawat, *op. cit.*

[53] *Ibid.*

[54] *Ibid.*

[55] *Ibid.*

[56] Fung, Doris, *op. cit.*

(TechTAS) in May 2018 for fast-track administrative arrangements for settling down foreign and mainland R&D talents.[57]

Funding emanates from the mainland central government as well. From mid-2018, the central government in Beijing have also made available science/tech funding to Hong Kong researchers and tertiary institutions through research remittance as part of the overall plans to strengthen the domestic tech sector, encourage tech development and cooperation with the Mainland for its main tech objectives.[58] Hong Kong's tertiary institutions enjoy strong rankings[59] in the QS University and Times Higher Education Ranking systems in science and engineering empowered by more in-house R&D spending and staffing, more commercialization of research outcomes and industry research collaborations.[60] Some Hong Kong universities like Hong Kong University (HKU) and Chinese University of Hong Kong (CUHK) have also opened campuses in China as well. Externally, the Hong Kong authorities are also looking into attracting foreign human talents to Hong Kong.

John Lee Ka-chiu vowed to augment Hong Kong's innovation and tech industries and draw outside talents to the island economy through new policies while local legislators like Duncan Chiu (tech/innovation functional constituency in the Legislative Council) urged the authorities to reach out to HKEX-interested firms to relocate their R&D facilities and conveyor belts to Hong Kong.[61] To promote I&T talents and STEM education, the Hong Kong education system has been carrying out curriculum updating, professional training for instructors, large-scale learning activities, Global STEM Professorship Scheme from 2021 to assist tertiary institutions in luring global I&T scholars and their research collaborators for STEM

---

[57] *Ibid.*

[58] *Ibid.*

[59] [Electrical & Electronic Engineering rankings: HKUST (28) HKU (43) CUHK (66) PolyU (76) CityU (85); Computer Science & Information Systems: CUHK (26) HKUST (29) HKU (39) CityU (74) PolyU (92); Chemistry: HKUST (30) HKU (41) CUHK (81); Chemical Engineering: HKUST (40) HKU (54); Mathematics: HKUST (37) CUHK (41) HKU (56) CityU (88); Physics & Astronomy: HKUST (50) HKU (67); Medicine: CUHK (29) HKU (40) in QS World University Ranking by Subject 2021. Source: Fung, Doris, *op. cit.*]

[60] Fung, Doris, *op. cit.*

[61] Zhang, Tianyuan, *op. cit.*

teaching and research.[62] Hong Kong's cosmopolitan welcome to external human talents has seen some foreign start-ups founded in the SAR with the 2021 InvestHK start-up survey indicating they made up approximately 28% of all start-ups, alongside multinational companies (MNCs) basing their R&D staff members in Hong Kong (with perceptions of a more familiar career and living environment) to service the Chinese market.[63]

The Chinese Transport Tech Sector and Foreign MNCs: Leading technological companies like IBM constructed a "smart" railway-management system for the SOE metro in Guangzhou while GE aviation technology worked with Aviation Industry Corporation of China in 2009 to build commercial planes.[64] Multinationals like these have the strongest hand in working with authorities when they have a unique leading technology that China desires with straightforward access to the top leadership.[65]

Another school of thought on the development of emerging cutting-edge technologies is that its development is most successful in the largest and most demanding clients (i.e. those in China) who are prepared to provide local clients with home-ground advantages. The Chinese authorities' decisions to backpedal on market sectoral access for foreign MNCs pressured the CEOs (Chief Executive Officers) of those companies to decide if they want to work with the authorities or lose access to a large consumer market. The Chinese three-way scheme to give their companies an edge over non-Chinese firms in cutting-edge technologies include features like the following:

(1) having the state as both buyer and seller in priority industries (e.g. state-owned CSR and China Railways, AVIC and China Eastern Airlines exercise state control over equipment acquisition, sales and development), (In the beginning of the 2000s, Alstom (France's TGV train), Kawasaki (Japan's shinkansen trains) and Siemens (German engineering) occupied 2/3 of the Chinese High-Speed Rail (HSR) market and they subcontracted the manufacture of basic parts to state-owned enterprises (SOEs) and gave end-to-end systems to Chinese railway operators. By early 2009, the state necessitated foreign MNCs

---

[62] Fung, Doris, *op. cit.*
[63] *Ibid.*
[64] Hout, Thomas and Pankaj Ghemawat, *op. cit.*
[65] *Ibid.*

involved in HSR initiatives to form joint ventures (JVs) with SOEs CSR and CNR where MNCs owned 49% stake in the JVs and provided their newest designs for manufacturing in China domestically (approximately 70% of each of these projects). Eventually, the SOEs became formidable rivals to the MNCs and recaptured their domestic markets. Having captured the domestic market, CSR and CNR manufactured core technologies in the approximately US$110 billion global rolling-stock market for high-speed rail, especially in developing/emerging economies for project funded by the Chinese, e.g. Belt and Road or BRI infrastructure) and even developed countries like CNR in Australia and New Zealand.)

(2) merging companies into a handful of national champions for economies of scale and consolidated learning (e.g. CSR and AVIC are results of mergers) from smaller-scale money-losing companies

(3) compelling MNCs to go into joint ventures (JVs) with their national champions for tech transfer in present and upcoming business deals and those who refuse to do so are marginalized.[66]

(*Source*: Hout, Thomas and Pankaj Ghemawat, "China vs the World: Whose Technology Is It?" dated December 2010 in Harvard Business Review (HBR) [downloaded on 1 January 2022]. Available at https://hbr.org/2010/12/china-vs-the-world-whose-technology-is-it.)

Since 2006, the Chinese authorities tried to acquire technologies from foreign multinational companies in sectors like air transportation, power generation, high-speed rail (HSR), IT and electric vehicles (EVs) by restricting their investments and market access to sensitive items, relying on domestic components made there, incentivise transmission of tech to their SOEs through JVs.[67] The Chinese state also heightened competitive instincts between international competing companies when bidding for megaprojects in China with the objective of eliciting more technological transfers in areas that Chinese SOEs are lagging behind.[68]

Not all relationships with Western/G7 MNCs are adversarial. Shanghai Automotive Industry Corporation and Volkswagen's joint venture (JV) came up with a vehicle for sale in developing/emerging markets while Shanghai Auto worked with GM to sell cars to India in order to outcompete

---

[66] *Ibid.*

[67] *Ibid.*

[68] *Ibid.*

lower-cost newer players like (South) Korea Electric Power in the developing/emerging markets.[69] Cummins partnered with the biggest Chinese diesel engine manufacturer (Dongfeng Motor) in product manufacturing and R&D to build car models in China in a shorter time and access new clients like urbanized mass-transit companies and also export made-in-China products from their Chinese base to other markets.[70]

HKSAR, in the meantime, is specializing in R&D activities. Under the Government's Hong Kong R&D Centre Programme, at least four institutions were set up to manage and commercialize applied R&D in niche sectors and coordinate tech transfers:

(1) Automotive Parts and Accessory Systems R&D Centre (under the Hong Kong Productivity Council),
(2) Hong Kong (R&D) Centre for Information and Communications Technologies (under ASTRI),
(3) Hong Kong Research Institute of Textiles and Apparel (hosted by the Hong Kong Polytechnic University),
(4) Hong Kong R&D Centre for Logistics and Supply Chain Management Enabling Technologies (hosted by the University of Hong Kong, the Chinese University of Hong Kong and the Hong Kong University of Science and Technology).[71]

China and the Global Energy Market: The Chinese state occasionally coordinates growth acceleration plans with new regulations for foreign MNCs, e.g. from 1996–2005, foreign firms occupy 75% of the Chinese wind energy market but, as the state nurtured the sector through subsidies, bidders needed to increase their domestically made quota from 40% to 70% while imported parts were subjected to much higher tariffs.[72] When the domestic wind energy market took off, foreign MNCs could not keep up with the orders through their supply chains while the Chinese domestic producers with licensed technology from the smaller European turbine producers met the ramped-up demand and so, by 2009, Sinovel and

---

[69] *Ibid.*

[70] *Ibid.*

[71] Fung, Doris, *op. cit.*

[72] Hout, Thomas and Pankaj Ghemawat, *op. cit.*

Goldwind dominated 2/3 of the domestic market, shutting out foreign firms from 2005.[73]

Chinese firms also captured the international silicon wafer/solar panel industry owing to the large-scale expansion of private firms with access to cheap financing and affordable land (sometimes free) offered by provincial-level officials, allowing the businesses to build their factories and apartment blocks that contributed to the companies R&D and other expenses.[74] Overproduction by these companies beat down solar panel global prices by half in 2009–2010, decimating Western companies like the German *Q*-Cells (loss of 60% in 2009 from previous years) while Chinese producers dominated 95% of the solar panels industry from 2010 with Suntech, Yingli and JA Solar making up 50% and 33.3% of the German and American markets, respectively.[75]

Drive Towards Greater Self-Reliance: Starting from 2016 when the Trump administration began to take a stronger US position in economic position *vis-à-vis* China that eventually resulted in technological decoupling, China began to move towards self-reliance in high technologies. From the 14th FYP (2021–2025) unfurled in March 2021, China aims to increase technological content in various sectors and boost R&D expenditure by 7% yearly until late 2025.[76] The state was particularly anxious that the Chinese yuan's increase in strength would inevitably increase to a point that would make Chinese low-value-added tech components/finished products exports non-competitive, and then their production lines would transition to lower-cost economies in Southeast Asia (or even India).

The Chinese government hopes to boost its semiconductor manufacturing capacity by setting up a country-backed funding to invest in the sector while trying to attract investments for the fields of advanced manufacturing, digital platforms, computing and other high-tech applications.[77] Diversification, local standards for technologies and attracting foreign tech Joint Ventures are just three ways to build up China's domestic tech industries. The Chinese tech industry is striving towards diversification

---

[73] *Ibid.*

[74] *Ibid.*

[75] *Ibid.*

[76] Yau, Emerald, "The rise of China tech" dated 30 April 2021 in FTSE Russell (downloaded on 30 April 2021). Available at https://www.ftserussell.com/blogs/rise-china-tech.

[77] *Ibid.*

with more than 100 large- and mid-cap companies giving investors choices to reach into China's most important innovators spanning the entire upstream to downstream value chain, including hardware and software tech like cloud to infrastructure serving 1.4 billion local customers.[78]

China also instituted product standards, benchmarks and specs that compel non-Chinese software producers to come up with country-specific versions for China in order to bypass Western patents and royalty requirements, e.g. Chinese WAPI and TD-SCDMA's wireless and 3G mobile telephone standards did not internationalize but they gave Chinese companies an advantage to outcompete foreign rivals.[79] China apparently found more challenges in joint venturing with Western software development firms and it had no complete leverage over the developed world's international business leaders in this field as there are few SOEs that can match them.[80] Accessing natural resources and having companies transfer their technologies to China are features that are managed by China's foreign diplomatic corp. and industrial policy-making. The planners in Beijing needed more ways to draw them to joint venture in China.

Local users also generate a lot of data (the famous adage "data is the new oil") which is another asset accessible for China's tech giants that have emerged in prominence globally. The CCP and state are aware of how "winner takes all" outcome in data-driven economies enable tech giants to accumulate socioeconomic power that interferes with the one-party system's control over political stability and order.[81] China's tech giants are known popularly as BAT or Baidu, Alibaba and Tencent. A newer version of this acronym is BATX with the addition of Xiaomi. Baidu is the Chinese equivalent of the Google search engine, Alibaba is the Chinese e-commerce engine, Tencent focuses on social messaging and gaming while Xiaomi is the Chinese equivalent of Apple in making smartphones. Tencent (a pioneering tech giant along with BAT which is the acronym for Baidu Alibaba Tencent) made public the use of the now-popular messaging app WeChat in 2011 that soon grew into a multifunctional app with social media, entertainment and mobile payment

---

[78] *Ibid.*

[79] Hout, Thomas and Pankaj Ghemawat, *op. cit.*

[80] *Ibid.*

[81] Gray, Joanne and Yi Wang, *op. cit.*

capabilities reaching more than 1.2 billion monthly active users.[82] BATX tech giants absorb smaller would-be or direct rivals, e.g. in 2020, Tencent sunk money into 168 investments/mergers and acquisitions in both local and global firms while Alibaba invested in 44 firms, Baidu in 43 companies and Xiaomi in 70 of them.[83]

Other than BATX, other major Chinese tech companies are Huawei and ZTE in the tech hardware sector. Perhaps it is useful to examine their case studies here.

The Dynamic Duo (ZTE and Huawei) and their Iconic Statuses: ZTE and Huawei are the dynamic duo and iconic poster children of China's digital tech giants. They represented China's best opportunities for global branding, cutting-edge technologies and maybe even one day, self-reliance on indigenously developed technologies. They are also the champions of China's state-led partially marketized economy and the fruits of China's economic reforms started by former paramount leader Deng Xiaoping. Paramount Leader Deng, former Vice Premier Xi Zhongxun (current Chinese President Xi Jinping's father) and Huawei all share a commonality, which is their link with the city of Shenzhen.

When Deng first started his economic reforms (circa 1978), he experimented gradually with SEZs using cities as living laboratories for the reforms. One of the SEZs was the city of Shenzhen (established May 1980) which started receiving investments from Hong Kong and Taiwan (so-called "Greater China" sphere) and then benefiting from the growth of the Pearl River Delta region that was fast becoming the "world's factory." Pearl River Delta would later form large conurbations of industrial zones when the authorities encouraged regionalism for economy of scale, convenience of logistics and travel and a larger market size.

Shenzhen originated as a small fishing village which eventually grew in population, encouraged by official policy to develop the city into a high-tech hub and eventually, it would become China's Silicon Valley. It famously grew from 30,000 in the 1970s to more than 10 million by the mid-2010s. It was in this rapidly expanding city that tech giants and the likes of ZTE, Huawei and Tencent were born. On many occasions and in many contexts, Huawei was touted as the best candidate to lead China's brand recognition to the world, since it had a foothold in the

---

[82] Yau, Emerald, *op. cit.*
[83] Gray, Joanne and Yi Wang, *op. cit.*

communications systems of the developing world, Europe, Japan and Oceania at the peak of its golden age of global business expansion.

Many tech giants, including ZTE and Huawei, were also at the core of China's "Made in 2025" plan where the country planned to develop indigenous technologies for their own systems. The ambitious plan was made known to the rest of the world and became public knowledge in May 2015. It immediately drew a lot of attention, from stakeholders to rivals, from interested parties to aroused competitors. The initiative "Made in China 2025," if realized, would bring about less reliance on foreign sources of technologies, including those from the US that currently owns, develops and researches many of the world's most cutting-edge technologies. In December 2017, the Chinese Ministry of Industry and IT began to declare its plans of developing China's own indigenous optical chips.

China also had its own concerns about overreliance on US technology. The Chinese authorities also perceived such reliance as possible sources of American espionage on its crucial communication systems and infrastructure. In the past, former Chinese President Hu Jintao famously said that even he had to use Microsoft Windows operating system in April 2006 to none other than the founder of Microsoft himself, Bill Gates. In the recent past, Chinese companies urged its SOEs, government departments and sensitive units to switch over from using US-made servers and data management systems to indigenously developed Chinese systems. In 2014, the government compelled China's banks to switch from their IBM servers and utilized local Chinese-made machines instead.

To whither down Chinese appetite for US high-tech products, the Chinese digital tech giants also invested lavishly in cutting-edge R&D. Huawei, for example, has its own R&D division known as HiSilicon Technologies. This outfit is currently enjoying large research budgets to develop native and indigenous technologies that many expect would compete with the large chip manufacturers in the US. Intel is probably one of the US chipmakers that manufacture similar products. Chinese chipmakers imposed competitive pressures on their American counterparts.

ZTE's Global Emergence: ZTE began its existence as Zhongxing Semiconductor Co. Ltd in 1985. Its pioneers were associated with the Chinese Ministry of Aerospace industry. Starting off as a State Owned

Enterprise (SOE), it evolved into a publicly trade company first listing in Shenzhen in 1997 and then the company looked outwards into the global marketplace in 2004, starting with Hong Kong. In 2004, ZTE enjoyed Hong Kong's link to the West and global city status ("Asia's global city") when it was listed on the Hong Kong Stock Exchange (HKSE). The HKSAR was the platform from which ZTE launched its global operations. ZTE benefited from the boom in wireless network and mobile phone businesses. The company also cited the triumvirate of intense R&D spending, recruitment of human talents and establishment of overseas branch offices as its secret ingredients of success in the initial stage.

The mobile phone business made it a brand name within China and some other niche areas of the world, but in 2017, the company and the Chinese mass media noticed that ZTE was conspicuously absent from the US market and hoped to outreach into that consumer market. This would be the last major market in the global arena which it had not penetrated. ZTE's achievements made it a remarkable company. In 2017, at the peak of its industrial development, ZTE attained the global number one position in the WIPO ranking system for patents' applications list. At that time, ZTE was busy expanding into continental Africa where Chinese companies were already the top investors in that region. ZTE's global presence was a testimony to the intense emphasis that the company placed on R&D. ZTE was enjoying a boom era in 2017, expanding its footprints both in the developing regions (like Africa) and developed economies in East Asia and European Union. The economy of scale in its global operations became a distinct advantage for further growth.

ZTE became known for its network carrier businesses and consumer electronic devices. In 2017, it was doing roaring business in Spain, Germany, Australia, Canada and other developed economies. ZTE was fast expanding its wireless system in the European arena in 2017. ZTE was also pumping in funding in the three trendy areas of Internet of Things (IoT), 5G technologies which China was a leading producer at that point in time as well as cloud computing. Cloud computing is connected with IoT in the sense that it makes data storage affordable, convenient and possible, resolving the problem of warehousing large physical servers (for data storage) that take up space.

In other words, 2017 represented ZTE's coming-out party and its place in the sun in the digital tech world. In East Asia (in countries that have implemented Chinese digital systems like Thailand), Russia, Japan and of course the domestic market of China, ZTE began to roll out new high-tech product offerings. These markets became the playing field for its latest "next-gen" technologies, including the technologies named after and optical transmission hardware (the same kind of technologies that Huawei vowed to get into, especially in the aftermath of the Meng Wanzhou incident).

The Rise of Huawei: Huawei was founded by a former People's PLA engineer and, like some other founders of Chinese digital tech giants like Alibaba's Jack Ma, Founder Ren Zhengfei was also a standing member of the CCP. In its genesis, the company was founded on a shoestring budget of US$3000 in Shenzhen in 1987. Western analysts attributed Huawei's phenomenal growth to state support. It made its fortune as an Original Design Manufacturer (ODM), basically functioning as a contractor for custom-made telecommunication devices and infrastructures for other countries. Chief Executive Officer (CEO) Richard Yu transformed the company from a hardware manufacturer to a smartphone consumer electronics designer/retailer and software provider.

Huawei was a latecomer to the smartphone but caught up very quickly and soon, by 2010, it was able to launch its own smartphone in the open market. Its IDEOS smartphone was a success and ran on Google's Android system and eight years later, Huawei was no longer operating entry-level Android phones but had already embedded Artificial Intelligence (A.I.) and cutting-edge chips armed with killer apps into its mechanism. Its smartphone had come of age.

What made Huawei formidable is that, like Apple, it designs its own Artificial Intelligence chips and need not rely on others. ZTE was behind Huawei in such capabilities. By mid-2018, Huawei was the second largest smartphone vendor/retailer in the world, overtaking Apple. Huawei aimed high in 2018, declaring to the American mass media that they have a target of overtaking Samsung in 2020 to be the number one smartphone vendor in the world.

Huawei's achievements were eclipsed by accusations of espionage, particularly from the West. The US had disallowed Huawei equipment into the country. AT&T backed out of selling Huawei telecommunication products in early 2018. Huawei's reputation was further tarnished when Australia banned Huawei 5G equipment from going into the country. The United Kingdom intelligence agencies released statements that they were unable to guarantee the reliability of Huawei equipment in the context of national security.

But, even the US alliance networks had some minor differential views when it came to Huawei's equipment. Canada and other Five Eyes (USA, Australia, Canada, United Kingdom and New Zealand) intelligence group members cautioned against using Huawei equipment. New Zealand's intelligence agency Government Communications Security Bureau banned use of Huawei technologies for the Kiwi nation's 5G network in February 2019 but may not rule out some form of non-essential role later on. The ban ignored a Huawei campaign that they were as essential to the telecommunications industry as All-Blacks to the game of rugby. Canada is now considering the same actions too, further complicated by the ongoing disputes over the detention of Meng Wanzhou. Flagship companies in US Alliance network countries/partners including Japan's Softbank began to take down Huawei hardware from their communication infrastructure.

In the Munich security conference of 2019 (15–17 February 2019) however, a rare disagreement broke out between US allies. United Kingdom was convinced that Huawei equipment should be removed from the core British communication system but its peripheral equipment like towers can still be utilized. Germany subsequently took a similar position. They soon found themselves pitted against the US position that all Huawei equipment posed risks to the security architecture within the US security network. US expected all allies to be responsible for the integrity and resilience of the Alliance's security and communication infrastructures.

Security Risks Cited by Intelligence Agencies: ZTE was punished for transgressing international sanctions placed on Iran for attempting to develop nuclear weapons programmes. In March 2018, ZTE admitted that it had sold technologies to Iran and North Korea despite the United Nations sanctions imposed against such activities. The US was further infuriated when, through its own informal sources/investigations, the US authorities found that ZTE punished some of the employees for the violations but awarded monetary incentives/bonuses to others who were equally culpable. This was the last straw for the Trump administration. US retaliation was fast and furious.

ZTE experienced technological ban from the US in April 2018, a ban that was lifted only in July 2018. The US Trump administration stopped Qualcomm chips and Google mobile phone OS (operating system) from being installed into ZTE systems and this decision immediately crippled ZTE product systems. It was a mortal blow to the company and exposed its soft underbelly when it comes to indigenously developed core technologies. In the aftermath of the ban, mutual fund managers including those in China quickly reduced the value of ZTE stocks in their portfolios. By July 2018, ZTE management was self-reflective and promised to learn from their past errors and strengthened compliance.

The Pentagon instituted its own policy of disallowing its personnel and department to sell ZTE and Huawei phones, accusing them of equipment used for spying on the US. This directive was put in place and implemented in May 2018. Eventually, the Pentagon sent the message out to retailers not to carry Huawei or ZTE phones. This sounded another death knell to the two companies' attempts to expand or roll out new products.

The US Department of Commerce also fined ZTE US$1 billion for the violations. ZTE also needed to apportion another US$400 million in funds to standby for any future violations of the international community's sanctions against Iran and North Korea. ZTE must also agree to inspections by the US of its factory facilities, supply chain as well as reshuffle its senior executives. Interestingly, the ban hurt ZTE so much that the company was literally on the verge of collapse.

It took Chinese President Xi Jinping's conversations with US President Trump before the latter did a U-turn on the ban to prevent too many ZTE jobs from being lost. ZTE was at that point running a global operation with tens of thousands of workers working on its product lines. President Trump ordered the US Department of Commerce to help the Chinese company get back on its feet again. The Congressional Budget Office (CBO) noted that the US$1.3 billion fine funding from ZTE was useful for budgetary purposes, which otherwise required drawing from the coffers of taxpayers' contributions.

When the Trump administration made the move against China's ZTE, he was doing it at a great risk because the decision to ban ZTE from American supplies was not uniformly supported by his colleagues. Even important members of his Republican colleagues disagreed with the Trump administration in various aspects. Thereafter, his decision to reverse the ban attracted criticisms from another group of colleagues. The hardliners and the hawks especially were deeply concerned that ZTE was a security threat to the United States. They were not satisfied with the sudden decision to reverse the ban.

Senator Mario Rubio of the Republican Party, a factional leader and a former Presidential Candidate was one of the US leaders who led the fight against ZTE's ban reverse and fought for the restoration of the ban. Rubio had always been concerned about China and its commitment to democracy and human rights, in the past, met Joshua Wong, a charismatic young leader of the Occupy Central movement in Hong Kong. He also had reservations and criticisms of China's Confucius Institutes (CI), fearing propaganda dissemination.

Rubio was convinced that China was US's number one foe when it comes to espionage on the US and intellectual theft. He considered the Chinese communications industry as a leading entity in China's adversarial industrial/commercial tactics on the US. When the ban was reversed and not reinstated, Rubio was convinced that the US had been outmanoeuvred, perhaps by the Chinese authorities.

A strong critic of US President Trump, Democrat Senator Charles Schumer on the other hand attacked Trump's weakness in the face of a rival state. Schumer was also concerned about job losses (a traditional Democrat concern) from US competition with China, in addition to Rubio's cited espionage and intellectual theft. Rubio was joined by a bipartisan united front led by Democrat Senator Chuck Schumer who has of late teamed up with former Democrat leader Nancy Pelosi to exercise power and influence over the Democrat-majority Senate to resist Trump's initiatives, including building the wall to separate the US from Mexico. Schumer considered ZTE to be a national security threat. Both Rubio and Schumer wanted the seven-year ZTE ban back on again.

Rubio and his associates' pro-ZTE ban advocacy was eventually defeated by intense lobbying efforts from Trump's Secretary of Commerce. The pro-ZTE ban factions eventually only agreed to disallow the Federal government to work with ZTE, Huawei or any of China's State Owned Enterprises (SOEs) and other commercial/industrial entities under the control of the Chinese government. The Democrats backed down from a seven-year-old ban. The White House lobbied and rallied hard to remove the ban. Secretary Wilbur Ross, successfully fought back against ZTE sanctions restoration.

Former President Trump's efforts put ZTE back into operational mode again was successful and it re-opened for stocks and shares trading in June 2019. Not surprisingly, its stocks plummeted sharply. ZTE's admission and capitulation to US charges indicated that company was in a much weaker position than Huawei as the latter had a much stronger retail network and technological base than the former. This point was made clear in a rare public conference by Huawei founder Ren who defiantly said that Huawei had the means to make up for supplies cut from the US and reckoned that Chinese indigenous production of replacement chips was not something in US interest.

But, even Huawei did not expect the unexpected. Before the ink was dry on the ZTE episode, a second front opened up in the digital competition between the US and China. Washington D.C. accused Huawei and its Chief Finance Officer (CFO) of deliberately conspiring to violate export laws to Iran and then hiding the tracks' evidence in its actions. The US Department of Justice also charged Huawei of illegally getting hold of American telecommunication company T-mobile's robot blueprint.

The US authorities then requested the extradition of Huawei's CFO from Canada. The Canadian authorities, acting upon that request, detained Meng Wanzhou and prevented her from leaving the country. Meng and her supporters obtained bail and she was allowed to leave detention with an electronic tag tied to her leg to prevent her from escaping. Meng had restricted free movement in accordance with bail conditions.

The daughter of the founder of Huawei, Meng Wanzhou, was detained in Canada at the US request for possible violations of sanctions against Iran. Canada complied and detained Meng and it sparked off a major row between Beijing and Ottawa. In fact, Canada's ambassador to China, John McCallum offered a way out by pointing out what he thinks were weaknesses in the US extradition treaty for Meng and went as far as to argue for an optimal situation in which the US drops the extradition treaty.[84] Five days later, McCallum was asked to step down by his Prime Minister Justin Trudeau. It was pointed out in the international media that former Canadian defence minister McCallum's wife was an ethnic Chinese and his home constituency had a large number of Chinese Canadians' residents in Ontario.[85] His comments appeared to be eminently inappropriate in the tense context of the Ottawa-Beijing standoff.

The magnitude of Beijing's response was not entirely anticipated by Ottawa. Shortly after Meng's detention, the Chinese authorities detained four Canadians (a diplomat on No Pay Leave, a businessman, a teacher

---

[84]BBC News, "Trudeau fires Canada's ambassador to China amid Huawei controversy" dated 27 January 2019 in BBC News (downloaded on 27 January 2019). Available at https://www.bbc.com/news/world-us-canada-47015700.

[85]*Ibid.*

and a drug dealer/user). Michael Spavor, businessmen with ties to Pyongyang and operated in the Chinese border city of Dandong in trade with North Korea was also detained. They released one of the three, Canadian teacher Sarah McIver, eventually. But, the Canadian individuals still under detention were seen as China's bargaining chips in the digital contestation game.

Eventually, the Canadian government revealed that up to 13 Canadians were detained after Meng's arrest (in total 200 Canadians are detained in China, before and after Meng's detention).[86] It remains unclear if these detentions were related to the Meng detention. But 8 of the 13 were subsequently released though few details were disclosed by the Canadian government surrounding their circumstances.[87] Meng got on bail after a C$10 million (US$7.4 million) deposit, wearing an electronic tag and observing a curfew from 11 pm to 6 am but allowed to reside in her Vancouver Canadian properties (she has two good-sized landed properties) as her legal team prepares to fight the extradition.

The first Canadian to be detained, Michael Kovrig, was a diplomat on no pay leave (NPL) status. In the international media circle, he was identified as a retired diplomat. He was suspected of working for a civil society organization and non-government organization (NGO), International Crisis Group, that was not legally approved by China (the organization could operate only legally in the Hong Kong Special Administration Region HKSAR). His status as a diplomat on NPL resurfaced later in connection with the possibility of the detained individual enjoying diplomatic immunity.

Another detained Canadian individual, Robert Lloyd Schellenberg, was suspected of dealing in narcotics. He was initially given a 15-year sentence but was later re-trialled and given a death sentence. Some observers suspected political reasons behind his re-trial and argued that it was politically motivated. Canada's elite political establishment tried to explain to their Chinese counterparts that the detention of Meng was done in accordance with their law enforcement agreements with the US. Only the Canadian courts, and not the executive and/or legislature, can determine her fate. At the point of this writing, both Meng and the two

---

[86] Reuters, "Canada says 13 citizens detained in China since Huawei CFO arrest" dated 3 January 2019 in CNBC (downloaded on 3 January 2019). Available at https://www.cnbc.com/2019/01/04/canada-says-13-citizens-detained-in-china-since-huawei-cfo-arrest.html.
[87] *Ibid.*

Canadians detained in the aftermath of Meng's detention were all released as part of an exchange made possible by intense diplomatic representations involving Canada, China and the US. Besides actual and alleged Iranian sanction violations, China is also cited by the West for having breaches or violations of confidentiality and privacy of clients' data. Regulations in China stipulated that the government and/or the CCP are able to demand the handover of data from the tech companies upon request.

ZTE, Huawei and Chinese Responses to the Concerns: ZTE's response to the US was apologetic and made a promise to punish employees who were violating the sanctions. The fact that ZTE was crippled so easily by a simple supply decision from US tech providers revealed the soft underbelly of China's technological base. Ironically, for some observers, this may have driven China to accelerate or at least carry on with "Made in China 2025." But, they soon discovered that technologies alone could not make a successful product. It took two hands (users and Huawei/ZTE) to clap and this realization came with the full-blown impact of US moves against Huawei. For Huawei and ZTE equipment to succeed, they need to convince regulators and users in other countries that their hardware and software were reliable, technologically sound, well-designed, reasonably priced and most importantly, do not pose a security threat to the user country. The charges of espionage went against the grain.

Huawei Chief Executive Officer (CEO) and founder Ren Zhengfei warned that if his company faces technological ban from the US, like the one that handicapped ZTE, his company would develop their own indigenous technologies. Ren also informed the international media that his daughter, Meng Wanzhou, would not be the successor to his business, an attempt perhaps to downplay the impact of her arrest on the company's future. Ren mentioned this even before the detention incident in Canada.

The ZTE and Huawei episodes indicated that the digital contestation between the US and China was now in full blow and it is a high-stakes electronic Great Game. The future of the two companies is up for speculation. There are contradicting evidence on how they are doing. In the first quarterly results in the aftermath of the US technological ban, ZTE indicated first quarter record results, but the full impact of the US ban was not in by then.

The ability of ZTE or Huawei to stay at number one or even a leading contender in smartphone technologies will depend on whether they can keep up with the latest trends. The playing field will include the rollout of the latest technologies that include the likes of 5G technologies, foldable phones and AR devices, which will be the next generation of technologies that all players will compete.

On 27 February 2019, Intel officially announced it is ending its relationship with Chinese chip manufacturer Unisoc. Intel and Unisoc were going to collaborate on 5G solution systems known as Spreadtrum and manufacture 5G phones and they would utilize Intel's XMM 8000-series modem with Spreadtrum's application processor that can be embedded into personal computers (PCs), mobile devices and smartphones.[88] Intel denies the decision is linked to the fact that Unisoc is a Chinese state-supported enterprise known as Tsinghua Unigroup (China's second-largest chipmaker) or part of the "containment" of Chinese high-tech industries.[89] Following the announcement of the end of partnership, Unisoc announced it would develop a 5G modem chip based on its own technologies.[90]

In 2015, the US government stopped Intel from supplying high-end Xeon chips to China for its supercomputer projects. The US prevented chip supplies to a Chinese-owned Inspur supercomputer supply firm and banned four supercomputer developers from receiving those chips: the National Supercomputing Center of Guangzhou, National University of Defense Technology (which built Tianhe-2 and Tianhe-1A, the world's fastest supercomputers at this moment) and two other entities which operated supercomputers allegedly to simulate nuclear explosions.[91]

New Priorities in the Tech Sector: In its next stage of tech development, China is financing large initiatives in emerging tech like next-gen nuclear

---

[88] Reichert, Corinne, "Intel ends Unisoc deal so it can work directly with Chinese OEMs" dated 27 February 2019 in ZDnet (downloaded on 27 February 2019). Available at https://www.zdnet.com/article/intel-ends-unisoc-deal-so-it-can-work-directly-with-chinese-oems/.

[89] Cheng, Ting-Fang, "Intel's 5G modem alliance with Beijing-backed chipmaker ends" dated 26 February 2019 in Nikkei Asian Review (downloaded on 26 February 2019). Available at https://asia.nikkei.com/Economy/Trade-war/Intel-s-5G-modem-alliance-with-Beijing-backed-chipmaker-ends.

[90] *Ibid.*

[91] Kan, Michael, "US blocks Intel from selling Xeon chips to Chinese supercomputer projects" dated 10 April 2015 in PCWorld (downloaded on 1 January 2019) in PCWorld. Available at https://www.pcworld.com/article/2908692/us-blocks-intel-from-selling-xeon-chips-to-chinese-supercomputer-projects.html.

reactors, nanotech, quantum physics, clean/green energy and potable water purification and they are also providing incentives or/and compelling multinational companies (MNCs) to co-share their technologies with Chinese SOEs through joint ventures.[92] Perhaps, the most relevant case study in this area is found in Industry 4.0-related industrial sectors. Hong Kong, for example, is keen to develop its Fintech sector as Fintech start-ups jumped from 138 (2016) to 472 (2021) as the global 8th top ecosystem in "Global Fintech Ecosystem Report 2020" released by Startup Genome and Hong Kong's cited advantages are as follows:

(1) a base for emerging/mature Fintechs in Hong Kong and Asia-Pacific where regional financial activities are found (for encouraging financial institutions to utilize Fintech, supervisory sandboxes established by HKMA, SFC IA facilitate financial institutions to accumulate actual big data and user feedback on Fintech products/services in a controlled environment),
(2) B2B Fintech emphasis, i.e. Fintech companies that service finance outfits' regional businesses,
(3) Luring of mainland Chinese Fintech, IT and e-commerce firms to Hong Kong as their regional and global operational hub. (Large mainland banks and insurance firms in Hong Kong facilitate access to mainland Fintech leaders too. To facilitate Hong Kong's transition to Smart Banking Era, the Hong Kong authorities are working hard. HKMA provided licenses to eight online banks to provide Internet-based retail banking services, permitted four virtual insurers in the insurance market to operate digital distribution channels and, in November 2020, instituted Commercial Data Interchange (consent-based financial facility for secure/efficient data flow between banks and data providers) and Central Bank Digital Currency for cross-boundary payments.)[93]

Since 2020, the fortunes of tech giants have changed due to the authorities' economic centralization at the macro level and stronger scrutiny of tech giants in China. China's digital policy has been modified in the Xi era to veer away from a focus on accelerated economic growth

---

[92]Hout, Thomas and Pankaj Ghemawat, *op. cit.*
[93]Fung, Doris, *op. cit.*

towards robust state control over the industry and closer adherence to traditional value system and Party agendas.[94] Even private-sector tech firms are only tolerated in China if they operate under state scrutiny, provide value-add to the national economy and must be prepared for incorporation or hybridization with state organizations.[95] In 2010, the state needed foreign software firms doing business with state-owned customers (SOEs) to provide access to their source codes (but they backtracked after strong complaints from the foreign companies and governments).[96]

In November 2020, an Ant Group (Alibaba) IPO did not go ahead and state regulators compelled the Ant Group to go for corporate restructuring the founder was questioned by the authorities.[97] Alibaba founder Jack Ma therefore has lied low due to the authorities' scrutiny of his company, including Alibaba's Ali Investment and Tencent's Literature Group receiving penalties of RMB 500,000 (about A$110,000) each for anticompetitive acquisition practices and contracts.[98]

In March 2021, Tencent and Baidu were penalized RMB 500,000 each for anti-competitive acquisitions and contractual arrangements and, in April 2021, the government summoned 34 platform firms (including Alibaba and Tencent) for "administrative guidance sessions" and then Alibaba and Tencent were penalized with RMB18.228 billion (around A$4 billion) and RMB500,000 fines, respectively.[99] In July 2021, the Chinese government stopped a merger to shore up Tencent's dominance in the gaming market.[100] In January 2022, President Xi Jinping promulgated tougher regulation and management of the Chinese digital economy to prevent "unhealthy" development and cease "platform monopoly and disorderly expansion of capital," perhaps essentially preventing excess accumulation of private sector capitalist power (e.g. monopolies) that may challenge the Party.[101]

After this chapter's survey on contemporary Chinese technological development, this volume proceeds to examine selected specific case

---

[94] Gray, Joanne and Yi Wang, *op. cit.*
[95] Hout, Thomas and Pankaj Ghemawat, *op. cit.*
[96] *Ibid.*
[97] Gray, Joanne and Yi Wang, *op. cit.*
[98] *Ibid.*
[99] *Ibid.*
[100] *Ibid.*
[101] *Ibid.*

studies in the following chapters, including how China powered its technological development in heavy industries and beyond through the use of coal energy.

Bibliography

BBC News, "Trudeau fires Canada's ambassador to China amid Huawei controversy" dated 27 January 2019 in BBC News (downloaded on 27 January 2019). Available at https://www.bbc.com/news/world-us-canada-47015700.

Charlton, Emma, "6 things to know about China's historic rise" dated 1 October 2019 in World Economic Forum (Weforum) (downloaded on 1 October 2019). Available at https://www.weforum.org/agenda/2019/10/china-economy-anniversary/.

Cheng, Ting-Fang, "Intel's 5G modem alliance with Beijing-backed chipmaker ends" dated 26 February 2019 in Nikkei Asian Review (downloaded on 26 February 2019). Available at https://asia.nikkei.com/Economy/Trade-war/Intel-s-5G-modem-alliance-with-Beijing-backed-chipmaker-ends.

Fung, Doris, "Innovation and Technology Industry in Hong Kong" dated 6 September 2022 in HKTDC Research (downloaded on 6 September 2022). Available at https://research.hktdc.com/en/article/MzEzOTIwMDIy.

Gray, Joanne and Yi Wang, "China's big tech problem: even in a state-managed economy, digital companies grow too powerful" dated 13 July 2022 in The Conversation (downloaded on 13 July 2022). Available at https://theconversation.com/chinas-big-tech-problem-even-in-a-state-managed-economy-digital-companies-grow-too-powerful-186722.

GT staff reporters, "China to accelerate building of intelligent, unmanned ports in next five years" dated 24 September 2021 in Global Times (downloaded on 24 September 2021). Available at https://www.globaltimes.cn/page/202109/1235005.shtml?id=11.

Han, Peng and Gao Yun, "'Ghost port': Asia's first fully-automated port begins operations in Qingdao" dated 13 May 2017 in CGTN (downloaded on 13 May 2017). Available at https://news.cgtn.com/news/3d637a4e31677a4d/share_p.html.

Hawksford, "Why Is Hong Kong a Unique Place for Tech Companies?" dated 7 April 2021 in GuideMeHongKong Hawksford (downloaded on 7 April 2021). Available at https://www.guidemehongkong.com/in-the-news/2021---why-is-hong-kong-a-unique-place-for-tech-companies.

Hout, Thomas and Pankaj Ghemawat, "China vs the World: Whose Technology Is It?" dated December 2010 in Harvard Business Review (HBR) (downloaded on 1 January 2022). Available at https://hbr.org/2010/12/china-vs-the-world-whose-technology-is-it.

Kan, Michael, "US blocks Intel from selling Xeon chips to Chinese supercomputer projects" dated 10 April 2015 in PCWorld (downloaded on 1 January 2019) in PCWorld. Available at https://www.pcworld.com/article/2908692/us-blocks-intel-from-selling-xeon-chips-to-chinese-supercomputer-projects.html.

Kobayashi, Shigeo, Jia Baobo and Junya Sano, "The "Three Reforms" in China: Progress and Outlook" dated September 1999 in Japan Research Institute (JRI) (downloaded on 1 January 2021). Available at https://www.jri.co.jp/english/periodical/rim/1999/RIMe199904threereforms/.

Reichert, Corinne, "Intel ends Unisoc deal so it can work directly with Chinese OEMs" dated 27 February 2019 in ZDnet (downloaded on 27 February 2019). Available at https://www.zdnet.com/article/intel-ends-unisoc-deal-so-it-can-work-directly-with-chinese-oems/.

Reuters, "Canada says 13 citizens detained in China since Huawei CFO arrest" dated 3 January 2019 in CNBC (downloaded on 3 January 2019). Available at https://www.cnbc.com/2019/01/04/canada-says-13-citizens-detained-in-china-since-huawei-cfo-arrest.html.

Yau, Emerald, "The rise of China tech" dated 30 April 2021 in FTSE Russell (downloaded on 30 April 2021). Available at https://www.ftserussell.com/blogs/rise-china-tech.

Zhang, Tianyuan, "HK to facilitate high-tech floats" dated 21 October 2022 in China Daily (downloaded on 21 October 2022). Available at https://www.chinadaily.com.cn/a/202210/21/WS6351f6c0a310fd2b29e7dbe9.html.

Chapter 3

Coal Companies, Their Technologies and Importance in China's Energy Mix[*]

To examine how China powered its contemporary technological development, it may be useful to start with the hydrocarbon energy source that it utilized most during its post-1949 industrialization. Chinese coal companies and their extraction technologies formed the backbone of energy provision in China.

Recent Developments in China's Coal Use: The single most important trend in the history of the global coal industry is the rise of demand from the PRC. China's use of coal has seen a spike of 392% from the year 2000 to hit 611.1 Mt in 2015 and in the process, expanded its share in global output from 26% (2000) to 56.1% (2015).[1] China greatly shapes global coal prices as well as supply and demand through its immense use of the resource. It uses coal to make 468 Mt of coke oven coke (66.0% of global volume), 823 Mt of crude steel (49.3% of international output), 712 Mt of pig iron (60.1% of global production) and approximately 2.50 Gt of cement (59.8% share of the world's production).[2] However, massive use

---

[*]Portions of this chapter is drawn from a limited-circulation Background Brief: Lim, Tai Wei, "Coal and its Importance in China's Energy Mix" dated 21 February 2018 in EAI Background Brief No. 1329 (Singapore: NUS EAI), 2018.

[1]International Energy Agency (IEA), Key Coal Trends Excerpt from: Coal Information 2016 (Paris: IEA), 2016, p. 5.

[2]Ibid., p. 9.

comes with its own set of challenges. Coal pollutes compared to other fossil fuels like oil and natural gas. To mitigate pollution, China is following a regional trend of shutting down small-scale coal-powered generators in favour of larger and more efficient power generation plants.

Historical Use of Coal in China: Japan, China and South Korea are located in a region where, with the exception of Japan and South Korea, the rest of the region is coal-rich or has sizeable deposits of coal. In China, the regions of Shanxi, Inner Mongolia and Shaanxi have been some of the largest suppliers of domestic Chinese coal. In terms of variety, the major genres of coal are lignite in east Siberia, the Far East region of Russia and Mongolia while the rest of the Northeast Asia region has mostly bituminous and sub-bituminous coal with some anthracite.[3]

Traditionally, heavy industrial demand for coal from sectors like steelmaking has been a major source of coal consumption in the Northeast Asian region. Electricity generation is another guzzler of coal fuels in the same region. More recently, coal use in China reached 3.97 billion tons (physical tons) in 2015 and 3.78 billion tons in 2016.[4] Heavy coal use for electricity, heating and industries also comes with its own set of challenges. The population density of Northeast Asian cities (especially first-tier cities in the region) meant that pollution was concentrated and dense in certain areas. This has led rising middle classes in those cities to express socio-political dissatisfaction, compelling regional and national governments to react and come up with policies to reduce pollution or to enforce penalties for anti-pollution laws.

China's growing need for coal will continue to drive its demand for domestic coal resources and even increase its imports. 70% of China's energy demand is met by its own coal resources (approximately 114 billion tons total verified basic reserves of coal) and growing imports from

[3]Fukushima, Atsushi, "Coal and Environmental Issues in Northeast Asia" dated March 2004 in The Institute of Energy Economics, Japan (IEEJ) website (downloaded on 1 January 2017). Available at http://eneken.ieej.or.jp/en/data/pdf/242.pdf, p. 9.

[4]Lin, Alvin, "Understanding China's New Mandatory 58% Coal Cap Target" dated 17 March 2017 in Natural Resources Defence Council (NRDC) website (downloaded on 17 March 2017). Available at https://www.nrdc.org/experts/alvin-lin/understanding-chinas-new-mandatory-58-coal-cap-target.

other countries.[5] China has continued to be the globe's largest coal producer since 1985 with a total production of 3.527.2 Mt of coal.[6] Currently, Qinhuangdao is the most important port for China's coal trade and the annual coal trade fair organized in the city in November 2016 was a facility for coal producers and electricity power plant managers to conclude more long-term deals overseen by the China National Coal Association.[7] From 1949, Qinhuangdao's major ice-free harbour facilities have been optimally utilized, exporting coal, coke, petroleum and timber and, in the case of coal, the port handles a large share of the overall national coal petroleum freightage.[8] Qinhuangdao, China's main coal port, saw China's coal imports increase 33% year on year (y/y) to 89.5 million tonnes in the first quarter of 2017 due to electrical power generation.[9]

Dalian for example has become a growing centre of coal trading for China. Due to its own increasing demand, China is the world's largest producer of coal as well, from prolific mines found in the northern as well as northwestern regions of China. Dalian was the location of the Fourth Northeast Asia Coal Trade Fair on 1 August 2013. Managers of approximately 600 firms that manufacture, or distribute coal were hosted by CNCA, China Coal Transportation and Sale Society and the Northeast

[5]Campi, Alicia, "The New Great Game in Northeast Asia: Potential Impact of Energy Mineral Development in Mongolia on China, Russia, Japan, and Korea" dated 2013 in the Asia-Pacific Policy Papers Series (downloaded on 1 January 2017). (Washington DC: Edwin O. Reischauer Centre for East Asian Studies), 2013. Available at http://www. reischauercenter.org/en/wp-content/uploads/2013/06/RC-Monograph-2013-Campi_The-New-Great-Game-in-Northeast-Asia.pdf, p. 12.

[6]International Energy Agency (IEA), *op. cit.*, p. 3.

[7]*Global Times*, "Annual coal trade fair in Qinhuangdao" dated 27 November 2016 in the Global Times website (downloaded on 1 January 2017). Available at http://www.globaltimes.cn/content/1020536.shtmlQinhuangdao.

[8]The Editors of Encyclopædia Britannica, "Qinhuangdao (Alternative Title: Ch'in-huang-tao)" in Encyclopaedia Britannica (downloaded on 1 August 2017). Available at https://www.britannica.com/place/QinhuangdaoChina's coal imports up 33% to nearly 90 million tonnes in Q1.

[9]Zeng, Xiaolin, "Coal mining in China" dated 30 May 2017 in the HS Fairplay: Maritime Shipping News website (downloaded on 30 May 2017). Available at https://fairplay.ihs.com/commerce/article/4287086/china-s-coal-imports-up-33-to-nearly-90-million-tonnes-in-q1.

Asia Coal trade centre.[10] Dalian has also become a major coal futures trading platform. The Dalian Commodity Exchange is now a premium platform for coking coal futures with more than one million contracts of iron ore futures transacting on a daily basis.[11]

The directives of the "Made in China 2025 blueprint" inked by the State Council in May 2015 to modernize/upgrade China's manufacturing sector through innovation and the use of IT in the manufacturing sector may impact the coal industry.[12] China is instituting efficiency increases by getting rid of wastes and obsolete heavy manufacturing technologies/activities, focusing coal utilization in big-scale efficient power and heat generation facilities while mothballing hundreds of thousands of inefficient, small-scale polluting coal-fired boilers.[13]

The "Energy Development Strategy Action Plan (2014–2020)" published by the State Council prescribes restrictions on annual primary energy and coal consumption until 2020 at absolute levels. It also sets quotas for lessening the coal share of primary energy consumption to less than 62% and increasing the percentage of the use of non-fossil energy to 15% by 2020 and to about 20% by 2030.[14] Therefore, the coal percentage of China's electricity generation is expected to go down from about 75% in 2012 to 45% in 2040; likewise the coal share of China's total primary energy consumption from 66% in 2012 to 44% in 2040.[15] Coal's percentage of total energy consumption dropped from 64.0% in 2015 to 62.0% in 2016.[16]

---

[10]Zhao, Qian and Roger Bradshaw (editors), "Dalian plays host to Northeast Asia Coal Trade Fair" dated 6 August 2013 in China Daily (downloaded on 1 January 2017). Available at http://www.chinadaily.com.cn/m/dalian/2013-08/06/content_16874898.htmEXCHANGES.

[11]Tan, Huileng, "China's wild futures trading is opening up a major opportunity for exchanges abroad" dated 21 May 2017 in the CNBC.com website (downloaded on 21 May 2017). Available at https://www.cnbc.com/2017/05/21/heres-how-commodity-exchanges-are-eyeing-chinas-wild-futures-trade.htm.

[12]US Energy Information Administration, "Chapter 4. Coal" dated 11 May 2016 in *International Energy Outlook 2016* (downloaded on 1 January 2017). Available at https://www.eia.gov/outlooks/ieo/coal.php.

[13]*Ibid.*

[14]*Ibid.*

[15]*Ibid.*

[16]Lin, Alvin, *op. cit.*

The Politics of Coal in China: Like most other resource industries in China, the coal industry has its own set of political considerations. Technocrats face the scrutiny of the party political leadership and may be affected by involvement in corruption cases or perceptions of not doing enough to solve coal industry problems. Under Chinese President Xi Jinping's anti-corruption campaign, in 2015, the Baoding City Intermediate People's Court sentenced Wei Pengyuan, former deputy chief of the coal bureau under the National Energy Administration, for accepting 211 million yuan ($32.5 million) in bribes and failing to account for 131 million yuan ($20.2 million) worth of missing assets.[17]

Even the development of coal mining cities is subjected to political scrutiny. For example, critics of Datong's excess infrastructure building claimed that the city's former mayor Geng Yanbo (nicknamed 'Demolition Geng', "Geng ChaiChai" or "Geng Smash-Smash") from 2008 to early 2013 was responsible for overbuilding of infrastructure including a 3.3 square km of "ancient city" replica that relocated 500,000 residents to make way for the project.[18] Geng's legacy has both supporters and detractors. Geng's critics like Renmin University economics Professor Tao Ran called his programme excessive but supporters of Geng considered him a visionary mayor and Director Zhou Hao even shot a Golden Horse-winning documentary titled 'The Chinese Mayor' based on Geng who was eventually promoted in 2013 to Taiyuan mayor in Shanxi's provincial capital.[19]

Besides corruption issues, infrastructure evaluations and political campaigns, socioeconomic effects and environmental impact of coal use have also become political fodder. Environmental political pressure on the coal industry also comes from the highest levels of government as Premier Li Keqiang publicly announced the leadership's intentions to 'make our skies blue again'.[20] On the other hand, coal stakeholders warned of the

---

[17] Associated Press, "Chinese Coal Official Who Hid $30M Admits to Corruption" dated 29 December 2015 in the VOA website (downloaded on 1 January 2017). Available at https://www.voanews.com/a/china-corruption/3122856.html.

[18] Zhou, Xin, "Decline and fall: the broken dreams of a Chinese coal-mining city struggling to address industrial overcapacity" dated 30 May 2017 in the *South China Morning Post* (downloaded on 30 May 2017). Available at http://www.scmp.com/news/china/economy/article/1935326/decline-and-fall-broken-dreams-chinese-coal-mining-city.

[19] *Ibid.*

[20] Bradshernov, Keith, "Despite Climate Change Vow, China Pushes to Dig More Coal" dated 29 November 2016 in the *New York Times* (downloaded on 1 January 2017).

social impact of closing down coal mines. In the NPC session in March 2016, Datong Coal Chairperson Zhang Youxi urged the government to look into the impact of projected large-scale retrenchments, citing how coalmines tend to hire 5,000–7,000 employees with approximately 20,000 dependent relatives, numerically equating mine closures with closures of cities and towns.[21]

In response to the top leadership's call and popular demands from an increasingly vocal middle class in 2016 for cleaner air and to deal with overcapacity, the National Development and Reform Commission (NDRC) further ceased the building of 50 gigawatts of coal power plants in 2016 and, in 2015, China closed down 290 million tonnes of outdated coal capacity.[22] In 2016, the Ministry of Environmental Protection stopped the use of trucks to transport coal at the port of Tianjin and closed the trucking route from Hebei's coal railway to ships in September 2016 in a bid to reduce coal supply and use in China. Sceptics however argue that ministries and industries at the local and provincial levels may dilute/ tamper with the effectiveness of legislation.[23]

Cutting down or increasing coal supply also comes with its own set of political-economic considerations. When coal supply was reduced in the fall of 2016 through various means like restricting trucking venues, prices of coal went up. However, when the restrictions are lifted to cope with demand, the increase in prices causes coal price to drop. In early 2017, drop in coal prices prompted senior management from 19 largest coal firms in China (including Shenhua) to lobby the government to re-introduced controls on coal output like the case in late 2016 after the peak demands of winter were over in China.[24] These 19 *de facto* coal industry sector lobbyists are also dialoguing with the state to re-introduce the limits on mines' output,

---

Available at https://www.nytimes.com/2016/11/29/business/energy-environment/china-coal-climate-change.html?mcubz=1.

[21]Zhou, Xin, *op. cit.*

[22]Meidan, Michael, "Political Transition Will Override China's Policy Targets" dated 17 March 2017 in the Chatham house website (downloaded on 17 March 2017). Available at https://www.chathamhouse.org/expert/comment/political-transition-will-override-china-s-policy-targets.

[23]*Ibid.*

[24]Tabeta Shunsuke, "China moving to renew coal mining curbs" dated 23 February 2017 in Nikkei Asia (downloaded on 23 February 2017). Available at https://asia.nikkei.com/Politics-Economy/Economy/China-moving-to-renew-coal-mining-curbs.

allowing mines to operate only for a maximum of 276 days annually.[25] The coal mining sector (mostly SOEs) in China fears an oversupply of coal will affect prices if no cuts are implemented; the state's priority on the other hand is price stability to ward off any potential social disturbance or dissatisfaction that can arise from inflation.[26]

Is the Coal Belt Declining? The Chinese state is reducing coal's contribution to its energy mix from 64.5% (2015) to 50% by 2030 to upgrade air quality and cope with overcapacity in coal and steel production as the need for electric power using coal fuels is also slowing down (coal prices have dropped by more than 50% between 2011 to 2016), alongside a maturing economy.[27] These measures designed to cope with excess overcapacity in steel and coal production have generated images and concerns of a declining coal belt, raising concerns about the impact it will have on the local economies of the affected coal mining towns.

Datong, a 1.7 million-strong city located 350 km west of Beijing and the base of the massive state-owned Datong Coal Mine Group with 175,000 workers, is representative of a stereotypical coal rust belt in China combating excess industrial capacity.[28] The city is facing impending large-scale retrenchments, unpaid salaries for miners, uncompleted infrastructures due to sagging demand, an underused multimillion-dollar sports stadium, high business taxes, property glut, risky financial situation, overreliance on coal and protests by thousands of coal mining workers in Shuangyashan bordering Russia.[29]

Datong's Yudong New District has witnessed abandoned high-rise buildings, an almost-finished 1.2 billion yuan sports stadium exposed to weather conditions, deserted library, opera house and fine-art museum and an inactive China Tang City; Datong's overall property investment declined by 41.4% in 2015 as the coal industry faced smaller demand.[30] However, to say that the coal industry faces an irreversible and absolute

---

[25] *Ibid.*

[26] Tabeta Shunsuke, "China moving to renew coal mining curbs" dated 23 February 2017 in Nikkei Asia (downloaded on 23 February 2017). Available at https://asia.nikkei.com/Politics-Economy/Economy/China-moving-to-renew-coal-mining-curbs.

[27] Zhou, Xin, *op. cit.*

[28] *Ibid.*

[29] *Ibid.*

[30] *Ibid.*

decline is not accurate as coal fuels will continue to be an important part of China's overall energy mix.

Coal Industry in China: Still Critical despite Its Tremendous Advance in Hydro and Solar Energy? Coal remains responsible for approximately 75% of China's electricity power as of late 2016 in spite of large hydroelectric dam constructions and the globe's biggest initiative to implement solar panels and construct wind turbine infrastructure.[31] China's National Action Plan for Air Pollution Prevention and Control's (NAPAPPC) mid-term review published on 5 July 2016 indicated eight major regions increased 50.8 GW of new coal-fired energy capacity from 2013 to 2015; in 2016, 42 GW of coal-fired capacity was work in progress and 11 GW of power was cleared for construction in 2015.[32] By comparison, from 2013 to 2015, only 10.8 GW of coal-fired capacity in eight provinces was reduced. Simply put, the increase in demand for coal-generated power outstripped the reduction in the same category, indicating the continued importance of coal as an energy resource.

Given that these investments are made by China, coal will continue to be a fuel of choice for Chinese power plants as coal power plants are long-term investments and each power plant can operate for 30–50 years.[33] China's continual coal power plant investment is to feed its increasing energy consumption. Dwarfing both Japan and South Korea, China is the biggest user of coal globally, consuming 76 quadrillion Btu of coal in 2012, making up 50% of global coal consumption.[34] Current trends are showing signs of slowing as China adjusts to the 'new normal' in economic

---

[31] Bradshernov, Keith, "Despite Climate Change Vow, China Pushes to Dig More Coal" dated 29 Nov 2016 in the New York Times (downloaded on 1 January 2017). Available at https://www.nytimes.com/2016/11/29/business/energy-environment/china-coal-climate-change.html?mcubz=1.

[32] Sherpard, Wade, "If China Is So Committed To Renewable Energy, Why Are So Many New Coal Plants Being Built?" dated 8 July 2016 in Forbes.com (downloaded on 8 July 2016). Available at https://www.forbes.com/sites/wadeshepard/2016/07/08/if-china-is-so-committed-to-renewable-energy-why-are-so-many-new-coal-plants-being-built/#36dbe12 35918.

[33] *Ibid.*

[34] Lim, Tai Wei, "The Coal Industry in the Northeast Asian Context" in Coal Mining Communities and Gentrification in Japan edited by Tai Wei Lim, Naoko Shimazaki, Yoshihisa Godo (Germany: Springer), 2019, p. 135.

development and as the energy industry goes through fundamental transformation to deal with politically sensitive environmental problems.[35]

The slowdown is visible in the most recent decline of 0.19 billion tons (physical tons) from 2015 to 2016.[36] Part of the slowdown is due to the impact of China's switching to renewable energy fuels. By 2020, more than 15% of China's energy is projected to come from non-fossil fuels such as wind, solar and hydropower while, ironically at the same time, it will also be the largest coal user in the world.[37] In the long-term scenario, however, China is investing and switching to renewable energy moving from polluting affordable coal to high-grade ultra-supercritical coal that can combust at very hot temperatures with a high level of efficiency, using some of the newest equipment in the world.

Trends in the Regional Coal Industry: Problems in coal use are well known. An example is acid rain from coal use in the region. Sulphur dioxide and nitrogen oxide are two by-products of coal use. They can cause respiratory problems and damage the ecosystem. Use of low-quality coal can worsen air pollution in the region. Densely populated cities found in the Northeast Asian region intensify the problem with the concentrated use of energy use. Heavy industries in areas like Shenyang or other regions producing cement and steel also add to pollution generated through coal use. Domestic household consumption of energy including coal is likely to go up in expanding urban cities. Driving urbanization will drive massive people from the agricultural areas in the countryside to urban cities.

Like China, other Northeast Asian economies are also switching to cleaner fuels. Environmental pollution is also a major universal challenge that faces coal use in Northeast Asia and a common solution in the region has been to phase out smaller, less regulated, family-owned (even illegal) coal plants that use obsolete technologies. The closure of smaller-scale coal-powered plants is a shift towards the utilization of larger plants. China is phasing out smaller-scale coal-fired plants in favour of large plants with capacities of 500,000 kilowatts.[38]

---

[35] US Energy Information Administration, *op. cit.*

[36] The author calculated this figure based on the data available at Lin, Alvin, *op. cit.*

[37] Sherpard, Wade, *op. cit.*

[38] Fukushima, Atsushi, "Coal and Environmental Issues in Northeast Asia" dated March 2004 in The Institute of Energy Economics, Japan (IEEJ) website (downloaded on 1 January 2017). Available at http://eneken.ieej.or.jp/en/data/pdf/242.pdf, p. 27.

There are regional environmental cooperation initiatives among Northeast Asian states like China, Japan, Russia, Mongolia and South Korea which have established formal diplomatic relations with each other after the Cold War. These countries realize the advantages of combining resources to tackle common trans-boundary and cross-border environmental problems and pollution (including those caused by coal use). In the early days of Northeast Asian cooperation, China had already expressed through a Senior Official's Meeting (SOM) in 1993 that Beijing is interested in interacting or dialoguing with Japan and South Korea on topics of technology transfer and financial assistance mechanism building. China had proposed that the three countries embrace the "spirit of Agenda 21" to embark on a joint research project on clean coal combustion technology (CCCT).[39]

The Chinese were less keen on joint research and monitoring systems for trans-boundary air pollution which are of greater interest to Korea and Japan.[40] However, cooperative initiatives are often affected by geopolitical rivalries, historical memories and territorial disputes. Working on common environmental problems including those caused by the use of coal fuel is useful for strengthening interdependence among Northeast Asian states since most of them are either highly industrialized (like Japan and South Korea) or are large emerging economies with maturing growth (e.g. China). Cooperation in such areas was however mired in political differences. Rapid industrial growth (and the accompanying urbanization) after China's economic reforms took off in the 1980s accentuated the use of coal for industrialization, heating and other purposes, contributing to regional pollution. This was the same case for Japan during her rapid industrial growth in the 1960s (the 'income-doubling' period).

China as a Regional Coal Industry Investor and Importer: The United Nations (UN) lists China as the world's biggest investor in renewable energy projects since at least 2011,[41] albeit with coal remaining critical

---

[39] Taek-Whan Han, "Northeast Asia Environmental Cooperation: Progress and Prospects" dated 23 September 1994 in the Nautilus Institute website (downloaded on 1 January 2017). Available at http://nautilus.org/trade-and-environment/northeast-asia-environmental-cooperation-progress-and-prospects-3/.

[40] Ibid.

[41] VOA, "Despite Renewable Energy Push, China Still Runs on Coal" dated 21 November 2011 in the VOA website (downloaded on 1 August 2017). Available at https://www.voanews.com/a/despite-renewable-energy-push-china-still-runs-on-coal-134331173/168226.html.

despite developments in renewable energies like hydropower and solar energy. Other measures to combat pollution issues have been adopted like conserving energy or efficient energy use to reduce wastage. To bring about lower energy consumption of energy and reduce pollution as well, the substitution of a much cleaner fossil fuel of natural gas and/or renewable energy resources like hydropower, wind, solar and geothermal; and turning to multiple sources of fossil fuels, including Australia, the Southeast Asian region and North Korea in the neighbouring region. Southeast Asia is also potentially a destination for Chinese coal investments as well.

Coal Trade and Politics between China and North Korea: Interestingly, North Korea has quite big reserves of coal approximated at 12 GT and numerous potentially undiscovered coal mines.[42] Its problems in coal extraction are manmade. Years of isolation, sanction and self-reliance *juche* have not brought about rapid economic development or cutting-edge industrialization. North Korea is an important supplier of coking coal for making steel in China.[43] Its low-cost labour potentially makes North Korean coal cheap but the payback for taking Pyongyang's resources is political repercussions. When the North Koreans tested a missile in early 2017, China turned back North Korean ships carrying coal for export. North Korea is under increasing international sanctions, even from its ally China, for its missile testing and nuclear programmes.

Its recent test fire of Inter-Continental Ballistic Missiles (ICBMs) and threats to attack Guam did not go down well with the international community. The US has promised tougher crackdown on Pyongyang's trade lifeline and financial activities while opening the door for talks. These sanctions and tougher stance against Pyongyang are likely to stifle North Korean industries (including coal mining) even further. There is an international dimension to the Sino-North Korean coal trade. On 5 August

[42] Streets, David G., "Northeast Asia coal trading center set up in Dalian" dated 1 July 1991 in the Glocom website (downloaded on 1 January 2017). Available at http://www.glocom. ac.jp/column/1991/07/energy_and_acid_rain_projectio.html, unpaginated.

[43] Reuters staff, "China says enforcing North Korea coal ban seriously, no violation" dated 21 April 2017 in Reuters (downloaded on 21 April 2017). Available at http://www.reuters. com/article/us-storm-harvey/trump-to-visit-victims-of-unprecedented-floods-in-texas-and-louisiana-idUSKCN1BD074.

2017 after the North Korean ICBM tests, the United Nations Security Council (UNSC) levelled new sanctions on Pyongyang with the support of China but China also informed the US not to engage in a trade war with China on other fronts. It is unclear if this was a Chinese bargain of going hard on North Korea to exact concessions in its own trading relationship with the US.

An important element in the international pressure on North Korea centres on coal trade and China has moved to stop the coal trade with North Korea even before the August 2017 UN sanctions. China makes up approximately 90% of North Korean trade and coal occupies 50% of overall North Korean exports and China had already suspended North Korean coal imports from Feb 2017 till the end of 2017.[44]

North Korea exports to China dropped to US$880 million between January to June 2017, which is a decline of 13% compared to the same months in 2016 and China's coal imports from North Korea declined sharply with a trickle of 2.7 million tons entering China in January-June 2017, a drop of 75% compared with 2016.[45]

Enforcement of Chinese coal ban is closely scrutinized by the international community. Reuters spotted a few fully loaded North Korean cargo vessels returning to North Korea on 11 April 2017 after the Chinese customs imposed the obligations under the UN sanctions on 7 April 2017 and informed Chinese trading companies to turn away North Korean coal shipments.[46] But, on Friday 21 April 2017, *NKNews.org* released news (which was picked up by the international media) that a few North Korean vessels were spotted in the vicinity of Tangshan northern Chinese port. Questions about such sightings triggered off Chinese Foreign Ministry spokesperson Lu Kang's response that China was "seriously enforcing" the ban on North Korean coal imports and that it may be possible that some North Korean mariners may "need to be looked after for humanitarian reasons," accounting for the presence of those ships.[47]

---

[44] Feng, Shirley, Yang Liu and Luna Lin, "China bans North Korean iron, lead, coal imports as part of U.N. sanctions" dated 14 August 2017 in The Washington Post (downloaded on https://www.washingtonpost.com/world/china-bans-north-korea-iron-lead-coal-imports-as-part-of-un-sanctions/2017/08/14/a0ce4cb0-80ca-11e7-82a4-920da1aeb507_story.html?utm_term=.2045009a70ea.

[45] *Ibid.*

[46] Reuters staff, *op. cit.*

[47] *Ibid.*

China's Coal Investments in Southeast Asia: A promising region for coal trade and business is Southeast Asia. One of the primary targets for such investments in Southeast Asia is Indonesia. Chinese SOEs and government loans teamed up to invest in Southeast Asian economies. While China is keen to shore up its credentials as the world's environmental power after the US backed off the Paris climate agreement and is actively adopting cleaner energy sources like renewables to lessen dependence on coal, its coal investments overseas have increased. At the point of this writing, Chinese financial institutions and enterprises are involved in a minimum of 79 coal electricity generation facilities/ projects adding up to 52 GW or more in electricity volume.[48] Such investment plans fit into Southeast Asian energy priorities because electricity needs are poised to go up by 83% from 2011 to 2035 and a large number of Southeast Asian countries are still using coal-powered electricity generation facilities.[49]

Having discussed mostly manufacturing industries up to this point in time, the next case study will focus on a non-manufacturing sector to examine how Chinese companies apply technologies in the agricultural sector.

Bibliography

Associated Press, "Chinese Coal Official Who Hid $30M Admits to Corruption" dated 29 December 2015 in the VOA website (downloaded on 1 January 2017). Available at https://www.voanews.com/a/china-corruption/3122856. html.

Bradshernov, Keith, "Despite Climate Change Vow, China Pushes to Dig More Coal" dated 29 November 2016 in the New York Times (downloaded on 1 January 2017). Available at https://www.nytimes.com/2016/11/29/ business/energy-environment/china-coal-climate-change.html?mcubz=1.

Campi, Alicia, "The New Great Game in Northeast Asia: Potential Impact of Energy Mineral Development in Mongolia on China, Russia, Japan, and

---

[48]Walker, Beth, "Chinese Investment Stokes Global Coal Growth" dated 24 September 2016 in The Diplomat (downloaded on 1 January 2017). Available at http://thediplomat. com/2016/09/chinese-investment-stokes-global-coal-growth/.

[49]Myllyvirta, Lauri, "Southeast Asia is planning 400 new coal power plants — what does that mean?" dated 13 January 2017 in the Energydesk Greenpeace website (downloaded on 13 January 2017). Available at http://energydesk.greenpeace.org/2017/01/13/southeast-asia-coal-plans-health-japan-indonesia/.

Korea" dated 2013 in the Asia-Pacific Policy Papers Series (downloaded on 1 January 2017). (Washington DC: Edwin O. Reischauer Centre for East Asian Studies), 2013. Available at http://www.reischauercenter.org/en/wp-content/uploads/2013/06/RC-Monograph-2013-Campi_The-New-Great-Game-in-Northeast-Asia.pdf.

Feng, Shirley, Yang Liu and Luna Lin, "China bans North Korean iron, lead, coal imports as part of U.N. sanctions" dated 14 August 2017 in The Washington Post (downloaded on https://www.washingtonpost.com/world/china-bans-north-korea-iron-lead-coal-imports-as-part-of-un-sanctions/2017/08/14/a0ce4cb0-80ca-11e7-82a4-920da1aeb507_story.html?utm_term=.2045009a70ea.

Fukushima, Atsushi, "Coal and Environmental Issues in Northeast Asia" dated March 2004 in The Institute of Energy Economics, Japan (IEEJ) website (downloaded on 1 January 2017). Available at http://eneken.ieej.or.jp/en/data/pdf/242.pdf.

Global Times, "Annual coal trade fair in Qinhuangdao" dated 27 November 2016 in the Global Times website (downloaded on 1 January 2017). Available at http://www.globaltimes.cn/content/1020536.shtmlQinhuangdao.

International Energy Agency (IEA), Key Coal Trends Excerpt from: Coal Information 2016 (Paris: IEA), 2016.

Lim, Tai Wei, "Coal and its Importance in China's Energy Mix" dated 21 February 2018 in EAI Background Brief No. 1329 (Singapore: NUS EAI), 2018.

Lin, Alvin, "Understanding China's New Mandatory 58% Coal Cap Target" dated 17 March 2017 in Natural Resources Defence Council (NRDC) website (downloaded on 17 March 2017). Available at https://www.nrdc.org/experts/alvin-lin/understanding-chinas-new-mandatory-58-coal-cap-target.

Meidan, Michael, "Political Transition Will Override China's Policy Targets" dated 17 March 2017 in the Chatham house website (downloaded on 17 March 2017). Available at https://www.chathamhouse.org/expert/comment/political-transition-will-override-china-s-policy-targets.

Myllyvirta, Lauri, "Southeast Asia is planning 400 new coal power plants — what does that mean?" dated 13 January 2017 in the Energydesk Greenpeace website (downloaded on 13 January 2017). Available at http://energydesk.greenpeace.org/2017/01/13/southeast-asia-coal-plans-health-japan-indonesia/.

Reuters Staff, "China says enforcing North Korea coal ban seriously, no violation" dated 21 April 2017 in Reuters (downloaded on 21 April 2017). Available at http://www.reuters.com/article/us-storm-harvey/trump-to-visit-victims-of-unprecedented-floods-in-texas-and-louisiana-idUSKCN1BD074.

Sherpard, Wade, "If China Is So Committed To Renewable Energy, Why Are So Many New Coal Plants Being Built?" dated 8 July 2016 in Forbes.com (downloaded on 8 July 2016). Available at https://www.forbes.com/sites/wadeshepard/2016/07/08/if-china-is-so-committed-to-renewable-energy-why-are-so-many-new-coal-plants-being-built/#36dbe1235918.

Streets, David G., "Northeast Asia coal trading center set up in Dalian" dated 1 July 1991 in the Glocom website (downloaded on 1 January 2017). Available at http://www.glocom.ac.jp/column/1991/07/energy_and_acid_rain_projectio.html, unpaginated.

Taek-Whan Han, "Northeast Asia Environmental Cooperation: Progress and Prospects" dated 23 September 1994 in the Nautilus Institute website (downloaded on 1 January 2017). Available at http://nautilus.org/trade-and-environment/northeast-asia-environmental-cooperation-progress-and-prospects-3/.

Tabeta Shunsuke, "China moving to renew coal mining curbs" dated 23 February 2017 in Nikkei Asia (downloaded on 23 February 2017). Available at https://asia.nikkei.com/Politics-Economy/Economy/China-moving-to-renew-coal-mining-curbs.

Tan, Huileng, "China's wild futures trading is opening up a major opportunity for exchanges abroad" dated 21 May 2017 in the CNBC.com website (downloaded on 21 May 2017). Available at https://www.cnbc.com/2017/05/21/heres-how-commodity-exchanges-are-eyeing-chinas-wild-futures-trade.htm.

The Editors of Encyclopædia Britannica, "Qinhuangdao (Alternative Title: Ch'in-huang-tao)" in Encyclopaedia Britannica (downloaded on 1 August 2017). Available at https://www.britannica.com/place/QinhuangdaoChina's coal imports up 33% to nearly 90 million tonnes in Q1.

US Energy Information Administration, "Chapter 4. Coal" dated 11 May 2016 in International Energy Outlook 2016 (downloaded on 1 January 2017). Available at https://www.eia.gov/outlooks/ieo/coal.php.

VOA, "Despite Renewable Energy Push, China Still Runs on Coal" dated 21 November 2011 in the VOA website (downloaded on 1 August 2017). Available at https://www.voanews.com/a/despite-renewable-energy-push-china-still-runs-on-coal-134331173/168226.html.

Walker, Beth, "Chinese Investment Stokes Global Coal Growth" dated 24 September 2016 in The Diplomat (downloaded on 1 January 2017). Available at http://thediplomat.com/2016/09/chinese-investment-stokes-global-coal-growth/.

Zeng, Xiaolin, "Coal mining in China" dated 30 May 2017 in the HS Fairplay: Maritime Shipping News website (downloaded on 30 May 2017). Available at https://fairplay.ihs.com/commerce/article/4287086/china-s-coal-imports-up-33-to-nearly-90-million-tonnes-in-q1.

Zhao, Qian and Roger Bradshaw (editors), "Dalian plays host to Northeast Asia Coal Trade Fair" dated 6 August 2013 in China Daily (downloaded on

1 January 2017). Available at http://www.chinadaily.com.cn/m/dalian/2013-08/06/content_16874898.htmEXCHANGES.

Zhou, Xin, "Decline and fall: the broken dreams of a Chinese coal-mining city struggling to address industrial overcapacity" dated 30 May 2017 in the South China Morning Post (downloaded on 30 May 2017). Available at http://www.scmp.com/news/china/economy/article/1935326/decline-and-fall-broken-dreams-chinese-coal-mining-city.

Part II

Case Studies of Tech Companies and Tech Development in China

Chapter 4

Technologies and Modernization in China's Farm Development

Approaches Through the Years to Meet Changing Objectives: China is supplying food to almost 20% of the global population with less than 1/10th of its farmland while catering to Western-style dietary tastes.[1] They consume almost three times the meat as in 1990 as well as more dairy and processed foods and so they had to venture abroad to buy farms in the United States, Ukraine, Tanzania and Chile.[2]

China has to adjust to its population's changing dietary tastes by purchasing farms to satisfy rising meat demand alongside disposable incomes of the expanding middle class, with the average Chinese consuming 24.4 kg of pork and 14 kg of chicken annually.[3] Consequently, between 2011 to 2021, Chinese agricultural firms have acquired farmlands and major agribusinesses like pork-processing major Smithfield Foods such that, at the beginning of 2020, Chinese owners owned approximately 192,000

---

[1] McMillan, Tracie, "How China Plans to Feed 1.4 Billion Growing Appetites" dated February 2018 in National Geographic (downloaded on 1 January 2021). Available at https://www.nationalgeographic.com/magazine/article/feeding-china-growing-appetite-food-industry-agriculture.

[2] *Ibid.*

[3] Low, Nelson, "China's Appetite for Meat is Still Growing" dated 16 November 2020 in Reuters (downloaded on 16 November 2020). Available at https://www.reuters.com/article/sponsored/china-appetite-still-growing.

agricultural acres in the US valued at US$1.9 billion, including farming, ranching and forestry lands.[4]

Besides changing dietary tastes for meat necessitating overseas ranch purchases, China is also seeking supply stability and greater self-reliance on grains to meet growing needs. In 2013, China took decisive measures to ensure its self-sufficiency rate by ensuring the preservation of sufficient arable land domestically and through overseas acquisitions or partnerships from 2014 onwards.[5] China has been looking at self-sufficiency in staple grains as a political ideology and a geopolitical strategy to break out of political isolation from hostile nations as President Xi Jinping instructed rural officials: "Our rice bowl should be mainly loaded with Chinese food."[6]

Self-sufficiency was also one of the most important goals in the historical past during the Maoist era when the Soviet-style command economy state instructed farmers what to farm and the products are handed over to the state, leading to distortions in market demand that led to a great famine in the late 1950s and early 1960s.[7] The collective system ended in 1981 when the state retained ownership of the land but distributed the rights to cultivate it equitably among villagers.[8] Economic and food-production reforms have enabled China's population to double its supply of daily calories as 50% of China's population joined the global middle class while hundreds of millions have moved out of poverty, cities have expanded by approximately 600 million people since 1980 with urbanites desiring better occupations and salaries in the industrial economy.[9]

Food security has always been the national strategy of the Chinese government, while its content has broadened over time. In the name of food security, the early farm reforms were carried out. Since the

---

[4]McCrimmon, Ryan, "China is buying up American farms. Washington wants to crack down" dated 19 July 2021 (downloaded on 19 July 2021). Available at https://www.politico.com/news/2021/07/19/china-buying-us-farms-foreign-purchase-499893.

[5]China Daily, "95% self-sufficiency urged for grains" dated 16 December 2013 in China Daily (downloaded on 1 January 2021). Available at https://www.chinadaily.com.cn/business/2013-12/26/content_17197186.htm.

[6]McMillan, Tracie, *op. cit.*

[7]*Ibid.*

[8]*Ibid.*

[9]*Ibid.*

establishment of the PRC in 1949, the extreme farmland revolution in the early 1950s took farmland from landlords and reallocated it to landless peasants with the objective of having farmers work on their own farmlands.[10] It was an age-old cherished historical aspiration, resulting in a large number of private smallholders but the second farmland collectivization reform in the mid-1950s reversed the smallholder plots into large collectives called the People's Commune with centralized control.[11]

These parameters placed limits on agricultural development in China in the direction of commercialization and big agri-businesses. It was the political and ideological objective at that time when the PRC was founded in 1949 to transform farmers' lives by redistributing land from rich to poor farmers and then nationalizing farmlands a few years later.[12] There have been various efforts during the early years to ensure self-sufficiency in grain, such as the collectivization of the 1950s and 1960s and de-collectivization and marketization since the late 1970s.

It was only in the late-1970s that China started rural reforms to establish a family-based contractual household responsibility system that bestowed farmers freedom of land use rights, decision-making and a meritocratic connection between rewards and output performance.[13] But, by the 1980s, the disparate small farm plots that resulted from farmlands allocated to individual households resulted in the retardation of the use of mechanical equipment and agricultural infrastructures[14] that were needed to scale up farm production and/or commercialize it. After 1979, China succeeded in attaining a very high degree of self-sufficiency in meeting food requirements and further enhanced food security through higher grain output made possible by technological adoption and breakthrough innovations and by 2018, the Chinese government began to shift towards

---

[10] Chen, Fu and John Davis, "Land reform in rural China since the mid-1980s" undated in Food and Agriculture Organization (FAO) (downloaded on 1 May 2021). Available at http://www.fao.org/3/x1372t/x1372t10.htm.

[11] *Ibid.*

[12] Johnson, Ian, "Barred From Owning Land, Rural Chinese Miss Spoils of Country's Success" dated 26 September 2019 in The New York Times (downloaded on 26 September 2019). Available at https://www.nytimes.com/2019/09/26/world/asia/china-land-rights-farming.html.

[13] Chen, Fu and John Davis, *op. cit.*

[14] *Ibid.*

increasing reliance on food imports in the context of domestic dietary changes.[15]

The second reason often cited in US scholarship by scholars like Mark Elvin argued that the Chinese agricultural economy and in general were trapped in a high-level equilibrium trap.[16] There was progress but stifled by a technological immobility that made any sustained qualitative economic progress impossible, where technological use was maxed out without the application of leap-frog scientific methods that can only be achieved by an industrial-scientific revolution.[17] Consequently, there is high per-unit land yield coupled with high costs of production known as the "high-equilibrium trap thesis" that attempted to explain China's stagnation and lagging behind the West in agri-business development and commercialization. This is coupled with the development strategy of stabilizing the rural population with land as fall-back insurance for peasant workers. The application of new farming technologies (especially the kind associated with mechanization) was also constrained by these factors.

In addition, the continuing challenge faced by China in the agricultural sector is that out of 334 million acres of arable land (the rest of China's land space is mainly mountainous or arid deserts), 37 million acres are polluted or in the process of restoration.[18] Chinese farms (broken up into 200 million farms) are simply not large-scale enough like those in the West to feed 1.4 billion individuals with Western-style diets.[19] Increasingly, China is facing not only challenges in self-sufficiency in grain but also other challenges. Fast growth, market-oriented industrialization and urbanization since 1978 have changed all that as China transforms into a global economic superpower with increasing wealth, technological sophistication and a growing middle class that wants better food supplies.

---

[15]Du, Jane, "China's Food Self-Sufficiency and Food Trade Dependency" dated 2 August 2018 in EAI Background Brief No. 1375 (Singapore: NUS East Asian Institute), 2018, p. 3.

[16]Little, Daniel, "The High-level Equilibrium Trap" undated in University of Michigan website (downloaded on 1 January 2021). Available at http://www-personal.umd.umich.edu/~delittle/elvin.pdf.

[17]*Ibid.*

[18]McMillan, Tracie, *op. cit.*

[19]*Ibid.*

By the 2010s, emphasis was placed on grain self-sufficiency because, in late 2013, Yuan Longping, "father of Chinese hybrid rice," indicated that China's grain crisis is looming, triggering off discussions within official circles on how to avoid a food crisis in the future.[20] On 5 March 2015, Premier Li Keqiang reiterated in his annual government report to the NPC and CCPCC 2015 the "three-dimensional rural issues concerning agriculture, countryside and peasantry" (*san nong wen ti*) and the nation's plan to maintain a minimal grain output of minimally 550 million tonnes to ensure food security.[21]

In the 14th FYP (2021–2025), China is trying to create agricultural growth by allocating more farmland space for cultivation and augmenting the crop harvests.[22] It is done through the use of technologies that improves crop and livestock management (within the context of arresting shrinking agricultural land due to urbanization) by demarcating a minimum area of 120 million hectares for farmlands in China.[23] The government stated in its 14th FYP that agricultural security in Chinese terms does not refer to ostracizing global imports but to maintain a minimum of 95% of the grain supply from domestic sources.[24] It has already achieved this while tapping into a "two markets and two resources" system drawing resources from domestic and overseas sources, something stated in its October 2019 white paper on food security.[25]

In addition to ensuring adequate supply of farmlands, China in recent years focused on technological solutions. In the 14th FYP, there is a clear emphasis on technological advancement and "security" in a broader sense, including food security. The 14th FYP highlighted the importance

---

[20] Du, Jane, "Food Security in China (II): The Evolution of Post-1978 Grain Policies" dated 23 April 2015 in EAI Background Brief No. 1019 (Singapore: NUS East Asian Institute), 2015, p. 1.

[21] Du, Jane, "Food Security in China (I): The Evolution of Post-1978 Grain Policies" dated 23 April 2015 in EAI Background Brief No. 1018 (Singapore: NUS East Asian Institute), 2015, p. 1.

[22] CGTN, "How can China improve its food security?" dated 7 March 2021 in CGTN (downloaded on 7 March 2021). Available at https://news.cgtn.com/news/2021-03-07/How-can-China-improve-its-food-security--YnGmalQX1C/index.html.

[23] *Ibid.*

[24] *Ibid.*

[25] *Ibid.*

of a secure supply of grain, cotton, sugar, meat and milk products, construction of a national food security industry belt, water-conserving irrigation facilities, strengthening R&D of large/medium-size smart machinery and increasing mechanization rate of crop cultivation/harvesting to 75%.[26] There are also constructions of demonstration zones in 300 counties with fully automated crop production, fully mechanized horticulture, full-scale automated breeding and deep soil conversion of hills/mountains into agricultural land.[27]

The 14th FYP will implement seed banks, agricultural superior seed technology research, bioengineered breeding and augmented meteorological services.[28] These research facilities include a national crop germplasm resource bank/nursery, national breeding/production facilities in Hainan, Gansu and Sichuan, a Heilongjiang soybean base and an expanded national germplasm resource bank for livestock, poultry and aquatic products.[29] The National Reference Laboratory for Animal Diseases and the Regional Center for Pathogeny Surveillance will have better animal vaccine cold storage, no-harm treatment facilities for dead animals, disease/pest epidemic monitoring stations, pesticide risk monitoring facilities and a forest/grassland pest control station.[30]

Beyond food security, China recognizes the importance of other aspects of rural development. In order to improve food security beyond self-sufficiency in grain security, the 14th FYP is also keen to improve the safety regulation of foodstuff by making them traceable through the food chain and this also means that production facilities like agricultural industrial parks, processing zones and modernization demonstration areas must be optimized.[31] The 14th FYP also noted storage of agricultural products needs to be high tech as well with the construction of 30 national and 70 regional agricultural product backbone cold chain logistics facilities,

---

[26]*Xinhua News Agency*, "Outline of the People's Republic of China 14th Five-Year Plan for National Economic and Social Development and Long-Range Objectives for 2035" dated 12 March 2021 in Georgetown University Center for Security and Emerging Technology (downloaded on 12 March 2021). Available at https://cset.georgetown.edu/wp-content/uploads/t0284_14th_Five_Year_Plan_EN.pdf, p. 55.

[27]*Ibid.*, p. 59.

[28]*Ibid.*, p. 55.

[29]*Ibid.*, p. 59.

[30]*Ibid.*, p. 55.

[31]*Ibid.*, p. 57.

improved preservation facilities of farmland-proximate markets and cold chain storage/transportation facilities of selected livestock/poultry slaughter and processing factories.[32]

Other than using technologies, it can be said that scientific rationalization is being augmented in China. Eastern Shandong farmer Huang Qinyong from the eco-agricultural demonstration field in Zhengbaotun town Xiajin County praised rationalization of farming techniques for his increased yields: "Farming has changed from manual work to technical work! ... the leaves of these corns are thick and broad. Under the guidance of experts, the temperature can be calculated accurately. When growing five days later than usual, the plants are 20 centimetres higher than the ordinary varieties."[33]

Rural education is also important for rural transformation, as seen in Chinese President Xi Jinping's rural vitalization programme. As part of the consolidation of the battle against poverty, Chinese President Xi Jinping directed in February 2021 the fine-tuning of the rural vitalization strategy promulgated at the 2017's 19th CPC National Congress to prioritize the modernization and revitalization of agriculture, rural areas and farmers' livelihoods in China.[34] Agricultural modernization was a key move towards the state's goals of building a modern socialist economy and country.[35]

In 2021, the goal of self-sufficiency has become even more pronounced. The objective of the yearly rural policy blueprint "No. 1 document" was re-focused on maintaining the stability of food supply coming from the pandemic-affected leading food-exporting countries in 2020 by upgrading domestic grain yields for 2021–2025.[36] In a February 2021

---

[32] *Ibid.*, pp. 57 and 59.

[33] People's Daily Online, "Chinese farmers take advantage of high tech to increase production" dated 17 August 2020 in People's Daily (downloaded on 17 August 2020). Available at http://en.people.cn/n3/2020/0817/c98649-9732362.html.

[34] Xinhua, "Xi Focus: Xi charts road map for rural vitalization after victory in poverty fight" dated 25 February 2021 in Xinhuanet (downloaded on 25 February 2021). Available at http://www.xinhuanet.com/english/2021-02/25/c_139766552.htm.

[35] *Ibid.*

[36] Patton, Dominique and Hallie Gu, "China steps up focus on food security in major policy document" dated 22 February 2021 in Reuters (downloaded on 22 February 2021). Available at https://www.reuters.com/article/china-agriculture-document-idINKBN2A M0H5.

press conference, Agriculture Minister Tang Renjian reiterated, because China's population was still growing, both Communist Party committees and local governments will have the common objective of constructing a "national food security industry belt" to link all the main grain areas and carry out important scientific projects in breeding and growing genetically modified (GM) crops.[37] Technological rationality in handling food sufficiency and security is key here.

Even with more food imports, China could keep up food self-sufficiency at nearly 98% (above the global 95% standard for national food security), but as Chinese consumers transition from basic staples to livestock products in more affluent diets, increasing livestock needs more resources for cultivation and may result in more environmental damage than vegan foods.[38] This is part of food security in China's central government's "No. 1 document" for 2021.[39]

Besides livestock, in the arena of crops, other reasons are pushing up food imports as well. Despite self-sufficiency for rice, wheat and corn at 95%, Chinese demand for food imports is rising due to free trade agreements with other countries reducing tariffs, greater demand for higher quality foods or similar quality but lower prices (attracting large distributors) and the availability of diverse sources like Brazil and Argentina that have become global giants in items like soybeans.[40] The objective of Chinese agricultural production is therefore to increase domestic supply while diversifying external trade.

The government viewed rural vitalization as a measure to consolidate poverty alleviation achievements and prosperity by increasing incomes and upgrading rural infrastructures/livelihoods while strengthening village grassroots governance to better implement central/regional

---

[37] *Ibid.*

[38] Wong, John and Chen Gang, "Climate Change Posing a Threat to China's Long-Term Food Security" dated 13 August 2014 in EAI Background Brief No. 945 (Singapore: NUS East Asian Institute), 2014, p. 3.

[39] GT Staff Reporters, "China to accelerate reform to agriculture, ensuring food security" dated 23 February 2021 in Global Times (downloaded on 23 February 2021). Available at https://www.globaltimes.cn/page/202102/1216300.shtml.

[40] Zhou, Zheng, "China needs to diversify agricultural imports, provide investors with opportunity" dated 20 October 2019 in the Global Times (downloaded on 20 October 2019). Available at https://www.globaltimes.cn/content/1167374.shtml.

governmental policies, and, for this purpose, President Xi inspected Maozhushan Village vineyard in April 2021 to understand how the grape industry revitalized the village within eight years.[41] There is also an element of ecological protection where President Xi himself inspected the ecological conditions and conservation progress of Lijiang River in April 2021 while advocating the elimination of previous developmental philosophy of developing the economy at the cost of the environment.[42]

Rural/agricultural Development in the Government's Agenda: Currently, the most important directive when it comes to agricultural development in China is the "No. 1 document." The "No. 1 document" 2021 has the objectives of bringing about a more self-reliant domestic seed industry, closer integration with manufacturing and services industries to diversify rural incomes of farmers, encourage revitalization of Chinese rural regions, emphasize grain security, augment resources centres for crops/poultry/livestock/marine fishery and provide long-term support to major breeding programmes.[43] On 21 February 2021, the State Council's annual rural policy blueprint "No. 1 document" pointed out that the CCP committees are working with local governments to increase provincial grain harvests, supporting its local seed industry in the name of post-pandemic food security for its 1.4 billion-strong population in the 2021–2025 period.[44] The document was released to the public by China's Ministry of Agriculture and Rural Affairs.[45] In the post-pandemic era, China hopes to have consistent output of soybeans, edible oilseed crops

---

[41] CGTN, "China moves to rural vitalization with focus on grassroots governance" dated 26 April 2021 in CGTN (downloaded on 26 April 2021). Available at https://news.cgtn.com/news/2021-04-25/President-Xi-Jinping-inspects-south-China-s-Guangxi-ZKMVrKjyeI/index.html.

[42] *Ibid.*

[43] Patton, Dominique and Hallie Gu, "UPDATE 3-China steps up focus on food security in major policy document" dated 22 February 2021 in Reuters (downloaded on 22 February 2021). Available at https://www.reuters.com/article/china-agriculture-document-idCNL1N2KR06C.

[44] *Ibid.*

[45] GT Staff Reporters, "China to accelerate reform to agriculture, ensuring food security" dated 23 February 2021 in Global Times (downloaded on 3 February 2021). Available at https://www.globaltimes.cn/page/202102/1216300.shtml.

like rapeseed and peanut to hedge against unable international supplies of edible oils and practice diversification of agricultural product imports while constructing a modern animal farming system and shoring up output capacity of pigs.[46]

The COVID-19 coronavirus pandemic is not far from the minds of the planners behind the No. 1 document. Jiao Shanwei, editor-in-chief of *cngrain.com* website focusing on grain news, noted that No. 1 document placed emphasis on the use of advanced technology and human capital to develop agriculture in the context of rising global food prices arising from pandemic-disrupted food supply chains, thereby necessitating self-reliance in grain supply.[47] It also appears that China is intending to produce national champions in food production through policy means and state assistance.

On 22 February 2021, in a media briefing, Vice Agriculture Minister Zhang Taolin opined that it was pertinent to choose superior enterprises for prioritized support alongside tighter enforcement of IPR in breeding and assistance for the ranking seed firms to set up private sector breeding systems.[48] Chinese seed companies like Beijing Dabeinong Technology Group Co, Shandong Denghai Seed and Winall Hi-tech Seed Co are all targeted in this effort to strengthen China's seed production and agricultural output.[49] Zhang Taolin, Vice Minister of the Ministry of Agriculture and Rural Affairs, also announced on 22 February 2021 that China's crop breeding acreage makes up more than 95% of the farming industry with trends towards domestically engineered seeds and a self-sufficiency ratio of key livestock, poultry and aquatic products have attained 75% and 85%, respectively.[50]

Li Guoxiang, a research fellow and agricultural/seed security specialist at the Chinese Academy of Social Sciences (a Ministry-level institution in China) decoded the main agendas of the No. 1 document 2021.[51] It is a document that promotes policy reforms in rural areas to industrialize the agricultural sector and bring about more manufacturing and services to increase agricultural production capacity by upgrading infrastructure and

---

[46] Patton, Dominique and Hallie Gu, *op. cit.*

[47] GT Staff Reporters, *op. cit.*

[48] Patton, Dominique and Hallie Gu, *op. cit.*

[49] *Ibid.*

[50] GT Staff Reporters, *op. cit.*

[51] *Ibid.*

upholding innovations in science and technology.[52] Li also mentioned that the diversification of rural industries is in more advanced stages in eastern China at the Yangtze River Delta region by processing more crops and fruits, increasing local farmers' incomes and attracting more farming talents to the region.[53] He hopes that the demonstrative showcase farm models can proliferate in the northern and western regions of China where incomes and living standards are relatively lower.[54]

Meeting Basic Supply Needs/Challenges Using Technology: China uses technology to meet the challenge of providing food supply to the biggest population in the world, a task of Malthusian proportion while ensuring food safety at the same time and maximizing the use of land with 20% of the world population with a mere 7% of its agriculturally usable land space.[55] Agricultural security in Chinese terms does not refer to ostracizing global imports but to maintaining a minimum 95% of the grain supply from domestic sources which it has already achieved while tapping into a "two markets and two resources" system drawing resources from domestic and overseas sources, something state in its October 2019 white paper on food security.[56] In this system, foreign food and agriculture exporters will continue to find a large Chinese demand while inviting foreign firms to share knowledge and skills to upgrade Chinese agricultural production.[57]

The Chinese government is also encouraging the industry by providing incentives for agricultural investments. The authorities provided a break on the rent and a 20-year contract for the overseas Chinese-founded Thai CP Group to change 6,425 acres of reclaimed mudflats outside Cixi City for agricultural activities with corporate social responsibility.[58] Wang Qingjun, a senior vice president of CP Group, is keen to have internal complementary in production, e.g. eggs that require growing grain for

---

[52] *Ibid.*

[53] *Ibid.*

[54] *Ibid.*

[55] Qin, Hengde, "Smart farming technology can transform Chinese agriculture and help feed the planet" dated 18 February 2021 in Global Times (downloaded on 18 February 2021). Available at https://www.globaltimes.cn/page/202102/1215818.shtml.

[56] CGTN, *op. cit.*

[57] *Ibid.*

[58] McMillan, Tracie, *op. cit.*

poultry feed, that in turn feed chickens to size as they lay their eggs before slaughtering and processing them once they are spent in laying eggs and then finally selling them in in-house grocery stores.[59] All these activities occur in-house within the company.

With the utilization of technologies, Chinese researchers point towards higher yields. Researcher Li Maosong from Chinese Academy of Agricultural Sciences highlighted the augmented livestock breeding, mechanization, crops' data and disaster prevention system as some of the ways to stabilize grain production.[60] The varieties of crops' contribution rate to agricultural yield has been boosted by more than 43% and the comprehensive mechanization rate of crop cultivation and harvest has gone beyond 70%.[61] There are demonstrative showcases for test-bedding technologies as well. Eastern Shandong farmer Huang Qinyong from the eco-agricultural demonstration field in Zhengbaotun town Xiajin County was enamoured with a new category of organic fertilizer and high-tech agricultural equipment to enable corn to sprout quicker and develop strong and firm roots; which had the capabilities of increasing harvest output of more than 300 mu (20 hectares) in 2020.[62]

It is useful to note China is not alone in upgrading grain value-addedness and food/agricultural diversity. Some argue that Northeast Asian economies such as Japan and Taiwan are also looking at similar high-yield agriculture in the context of high-cost production. In this context, the issue of farm subsidies may be relevant to the quest to upgrade the agricultural sector. The Council of Agriculture (COA) provides hundreds of millions of US dollars yearly subsidies to farmers who use their land to grow grains with higher economic value like corn, wheat, sugar cane, oil-seed camellia or non-genetically modified (non-GM) soybeans/flint corns or to farmers who implement environmentally friendly features while the Council rezoned more farmlands for agricultural development since 2018.[63]

---

[59] *Ibid.*

[60] People's Daily Online, *op. cit.*

[61] *Ibid.*

[62] *Ibid.*

[63] Lin, Chia-nan, "Farmers using land for agriculture to receive subsidies" dated 10 January 2018 in *Taipei Times*. Available at https://www.taipeitimes.com/News/taiwan/archives/2018/01/10/2003685522.

The same goes for Japan which also has ambitions in food/agricultural product diversification and agri-business exports. In 2021, the Suga administration revised Japan's Act on Special Measures to Facilitate Investment in Agricultural Corporations to increase special agricultural investments and facilitate the growth of financial institutions specialized in funding agricultural businesses as well as investments in the fishing, forestry and food-processing sectors.[64] This can help build up producers/exporters of agricultural and food products through expanded agricultural exports from 922.3 billion yen (US$8.5 billion) in 2020 to 2 trillion yen (US$18.5 billion) by 2025 and 5 trillion (US$46.1 billion) by 2030.[65] The sizable investments in infrastructure building, foreign marketing/branding exercises (e.g. through overseas food fairs organized by the Japan Food Product Overseas Promotion Center) and human resources development are nicknamed *seme no nosei* ("proactive agricultural policy") to feed increasing popular demand for Japanese food products in the world.[66]

Harnessing Industry 4.0 Technologies: Smart farming technologies such as drone, satellite imaging and pattern modelling, enable agriculturalists to tap into their smartphones and use intelligent environmental tools to guide their agricultural efforts, keep fertilizers/pesticides use to a minimal and conserve water.[67] Agriculture digital services enable product traceability, provide consumers' capabilities to scan QR codes for information on farmland location, harvest seasons/date and are environmental sustainability.[68]

Technological upgrading of agricultural production is a major theme of China's 14th FYP with the use of 5G technologies to manage crops and farm animals without massive manpower investments. Shepherds can monitor their flock using GPS-enabled apps, satellite sensing to reveal water sources and optimal volumes in their utilization while the IoT operates an AI-guided crop harvesting machine to chop the crops at the

---

[64] Sasada, Hironori, "Challenges in boosting Japan's agricultural exports" dated 12 March 2021 in East Asia Forum (downloaded on 12 March 2021). Available at https://www.eastasiaforum.org/2021/03/12/challenges-in-boosting-japans-agricultural-exports/.

[65] *Ibid.*

[66] *Ibid.*

[67] Qin, Hengde, *op. cit.*

[68] *Ibid.*

optimal heights to harvest the largest volume of grains.[69] Anhui farmer Yu Wenhe is harnessing technology to plant and harvest his consistently high-output crops that are impervious to the climate conditions and manipulated the whole process at home using smartphone apps, remarking smartphones have become farm tools to transplant, seed and fertilize the farmlands equipped with global positioning system for data management.[70]

Other Industry 4.0 technologies implemented in the Chinese agricultural sector include autonomous vehicles (AVs). The Chinese government is supporting local driverless tech firms with national trials under the auspices of industry group Telematics Industry Application Alliance (TIAA) that include SOE tractor manufacturer YTO Group, navigation systems developer Hwa Create and Zoomlion Heavy Industry Science & Technology Co. Ltd. that makes harvesters in collaboration with Jiangsu University.[71] Veteran companies like YTO first came up with their driverless tractor prototypes in 2017 while others like Lovol Heavy Industry Co. Ltd. are working with digital companies like Baidu since April 2019 to incorporate Baidu's Apollo automated driving system into its agricultural technologies.[72]

Since autumn 2018, driverless harvesters are going through trials in the rice fields, planting rice saplings, fertilizing, harvesting, chopping up golden rice stalks (and wheat and corn) and carrying loaded fertilizers.[73] This is part of a plan to have full mechanization by 2026 in order to mitigate the future ageing population and lack of young Chinese workers willing to do physical hard work on the farms.[74] The driverless tractor markets will need to adapt to the comparatively smaller sizes of Chinese farmlands as wheat farmer Li Guoyong from Hebei province noted that very bulky driverless tractors are not pragmatic given that more than 90% of Chinese farms are less than one hectare in size.[75] Smaller Chinese

---

[69] CGTN, *op. cit.*

[70] People's Daily Online, *op. cit.*

[71] Gu, Hallie and Dominique Patton, "On the autofarm: China turns to driverless tractors, combines to overhaul agriculture" dated 16 January 2019 in Reuters (downloaded on 16 January 2019). Available at https://www.reuters.com/article/us-china-farming-technology-idUSKCN1PA0DV.

[72] *Ibid.*

[73] *Ibid.*

[74] *Ibid.*

[75] *Ibid.*

farms may not need or be able to afford expensive regular-sized driverless tractors priced at approximately US$90,000 so the authorities are looking at land rights reforms to lease more farmland space to farmers.[76]

Foreign firms are contributing to China's high-tech development as well. Thai conglomerate CP Group has 3,600 acres of rice paddies (including 115 acres dedicated to organic rice while breeding crabs in them for food), greenhouses, broccoli fields, drones to disperse chemicals, a dumpling factory, a one-million-hen egg factory, temperature-sensitive robots to cull dead birds, recycled chicken manure to manufacture 22,000 tons of organic fertilizer per year.[77] It also owns airy/translucent six 30-foot vertical farms with rotating plant beds growing bok choy/amaranth/garlic chives, controlled environment for targeted fertilizer use and elimination of most pesticides while having four times the yield of a field with the same carbon footprint.[78]

Size Matters: Despite having mostly small farms, China has simultaneously developed some of the largest farms in the world such as the eight barns and processing plant at Modern Farming's Bengbu Farm in Anhui Province.[79] This is the largest dairy farm in China at almost 600 acres rearing 2,880 milking cows each while calves and pregnant cows are found in separate areas, with a cumulative total of 40,000 cattle (one of the largest in the world).[80] China is also open to more large-scale agricultural investments. China is encouraging global multinational companies (MNCs) to invest billions of yuan in agrifood complexes featuring fields, farms, factories, corporate offices and employee housing in the form of apartments to waterfront villas, something like what the Thai Chinese-founded CP Group is doing.[81]

An advantage of building large-scale industrial farms is that experts believe they can implement food safety and quality better than smaller farms. Scott Rozelle, an expert on rural China at Stanford University, argued that industrial dairies and slaughterhouses are favoured for making traceability and accountability possible for Chinese consumers, given that

---

[76] *Ibid.*
[77] McMillan, Tracie, *op. cit.*
[78] *Ibid.*
[79] *Ibid.*
[80] *Ibid.*
[81] *Ibid.*

"[The vast number of small farms makes China's food system] almost completely unmanageable in terms of food safety."[82] Food safety had been in focus since the 12th FYP (2011–2015) when the state set up national food safety standards to enforce food safety monitoring and the 13th FYP (2016–2020) reinforced this direction by formulating a national plan for food safety regulation.[83]

The Chinese authorities are very interested in implementing land trusteeships. This complements the concept of modern agribusinesses that taps into the economy of scale to churn out greater output production from large farms by merging and consolidating smaller farms while utilizing farmland trusteeship service organizations to take over and farm the lands from farmers who cannot or are unwilling to do so.[84] To satisfy the demand for better food and more Westernized diet and to meet its changing appetites with domestic crops, agricultural economist at Peking University Huang Jikun argued that irrigation must be improved, utilize greater automation and technologies and expanding the small-scale farmlands.[85]

Almost all the large-scale farms in China are operated by the government, cooperatives and businesses but sparks of entrepreneurship can be found among individual Chinese farmers. Individual farmers like Inner Mongolia's Liu Lin were inspired by automated American-style farming (who learnt about it through radio broadcasts) and sold the idea to the local authorities to lease 2,470 acres of farmland.[86] It is operated with American and European equipment that completed alfalfa harvesting tasks for industrial dairies in four hours (a task that would have taken 30 workers 20 days).[87]

At the same time, China does not want to match American farms in size because China's staple crops of corn, rice and wheat all yield the optimal amount of food per acre at modest-sized farms and consolidating large numbers of small farms may create social instability by resettling millions of farmers.[88] Instead, the Chinese authorities prefer to combine

---

[82] *Ibid.*

[83] Du, Jane, "China's Food Security and Food Safety" dated 25 August 2016 in EAI Background Brief No. 1166 (Singapore: NUS East Asian Institute), 2016, p. 3.

[84] CGTN, *op. cit.*

[85] McMillan, Tracie, *op. cit.*

[86] *Ibid.*

[87] *Ibid.*

[88] *Ibid.*

fields into farm clusters that have the same approximate area as a Walmart Supercenter parking lot.[89] The Chinese are cautious about placing over-reliance on large-scale farms. They are doubtful that farmers can work for the large industrial farms and rent out their own farmlands to get dual incomes as Ye Jingzhong, a rural sociologist at China Agricultural University in Beijing, explained that large-scale farms hired limited numbers of low-waged employees to maximize profits.[90]

Large industrial animal farms can also bring about environmental pollution and health hazards since high-volume production can generate waste and pollution (adding to existing pollution).[91] The Chinese authorities are keen to dispose of animal wastes sustainably, e.g. Bengbu-based Modern Farming utilizes a biogas digester to turn manure into energy to satisfy 33% of the facility's energy needs while using processed manure as fertilizers.[92] Thus, China and its authorities appear to accept only industrial farms only if they are environmentally friendly and can dispose of their waste products effectively.[93]

There is also an emergence of lifestyle urban farmers. Some wealthy urban dwellers have a lifestyle preference leaning away from reliance on industrial farming due to the trust factor. In northern Beijing, Jiang Zhengchao has a five-acre urban farm next to a highway with almost a hundred crops (watermelon, eggplant, taro, sweet corn, etc.).[94] The crops were sold at the distributor markets feeding and delivering fresh foods weekly to middle-class Beijingers who pay in six-month instalments while leasing plots to urbanites interested in farming and will help them take care of the crops for a fee.[95] Jiang is part of a movement by hundreds of countryside natives and university-educated Chinese youths tending to the farmlands known as "fanxiang qingnian" (young people going back to the rural areas).[96] They have started their own association known as Wotu Sustainable Agriculture Development Center and publish a magazine

---

[89] *Ibid.*

[90] *Ibid.*

[91] *Ibid.*

[92] *Ibid.*

[93] *Ibid.*

[94] *Ibid.*

[95] *Ibid.*

[96] *Ibid.*

known as "Sustainable Farming," riding on the back of China's organic boom with sales increasing at least 30 times since 2006.[97]

The CSA movement is also boosted by graduate student returnees who studied food activism in the US and they are eager to cater to urban working parents with limited time to cook and/or share scepticism about industrialized farms.[98] More than 122 community-supported agriculture (CSA) projects are facilitated by this movement, alongside a handful of Western-style farmers' markets in large cities catering to customers who trust their safe food grown amidst China's agricultural traditions.[99] Rural China scholar Wen Tiejun (and founder of model CSA Little Donkey since 2008) harked back to traditionalism by highlighting 40 centuries of agriculture in Asia with a dual emphasis on food sufficiency and an ideal natural environment at the same time.[100]

There is a whole ecosystem of stakeholders with funders, investors, governments, SOEs, private sector entities, universities and others. Even foreign academics and world-class consultants are included in this group. For example, the Thai CP Group has employed ranking American business scholars as well as world-class consulting firms such as McKinsey & Company to contribute to the farming projects.[101] Some investment analysts are quite bullish about the prospects. Alexious Lee, Head of China Industrial Research at Hong Kong brokerage CLSA, projected that China is likely to indigenize high technologies rapidly due to the Chinese corporations' ability to tap into China's own navigation satellite system "Beidou" to develop agri-businesses, an industry dependent on digital communications and global mapping systems.[102]

Agri-food investors appear to agree with the investment analyst's projections when it comes to high-tech farming in China. Even overseas Chinese conglomerates are scaling up their operations in China as well when investing in high-tech farms in China. At the Hangzhou Bay automated mega-farm (Asia's largest) owned by Thai Charoen Pokphand (CP) Group (animal feed conglomerate), three million hens produced approximately 2.4 million eggs daily with robots auto-detecting and

---

[97] *Ibid.*

[98] *Ibid.*

[99] *Ibid.*

[100] *Ibid.*

[101] *Ibid.*

[102] Gu, Hallie and Dominique Patton, *op. cit.*

discarding dead birds requiring only one employee to look after 168,000 chickens.[103]

AVs are another example of advanced technologies implemented in larger Chinese farms. Cheng Yue, general manager of tractor maker Changzhou Dongfeng CVT Co Ltd (that is trialling autonomous driverless vehicles in the rice paddies of Jiangsu province's Xinghua County) is confident that China's US$60 billion farming machinery industry and its world-class big data industry will enable consumers to know planting conditions, fertilizer/pesticide use when consuming agricultural products like rice.[104] Deputy Director Wei Xinhua of the School of Agriculture's equipment engineering department at Jiangsu University pointed out that sensors in driverless vehicles can also track crop conditions to adapt to different situations while becoming big data collectors on fertilizers volumes, assuring customers of quality produces with these data printed on food labels for the products.[105]

In some cases, the tech giants themselves are engaging in the farming and agricultural industries. China is already globally the largest pig farming industry making up 50% of the world's number of live pigs and AI technologies are being utilized to monitor diseases and track pigs while facial recognition software can identify pigs and their individual weight, diet regime and exercise.[106] The pork industry in China grew rapidly from farms (with 50 pigs each) making up 25% of the market in 2001 increasing to almost 75% by 2015 (the Shanghai wholesale meat market alone sells 2,500 pigs from midnight to sunrise per day).[107] It is now the world's biggest pork factory, Jinluo Meat in Linyi, which has 4,000 employees with an output of 267 products yielded from 32 million pigs acquired from independent farms.[108]

Chinese tech giants like Huawei, JD.com and Alibaba are collaborating with Chinese pig-rearing farmers to introduce new technologies with Huawei representatives articulating: "The pig farming is yet another

---

[103] McMillan, Tracie, *op. cit.*

[104] Gu, Hallie and Dominique Patton, *op. cit.*

[105] *Ibid.*

[106] Harper, Justin, "Huawei turns to pig farming as smartphone sales fall" dated 19 February 2021 in BBC News (downloaded on 19 February 2021). Available at https://www.bbc.com/news/business-56121470.

[107] McMillan, Tracie, *op. cit.*

[108] *Ibid.*

example of how we try to revitalize some traditional industries with ICT (Information and Communications Technology) technologies to create more value for the industries in the 5G era."[109] The Chinese authorities do believe that there are limitations to what technologies can achieve and have filled those gaps with policy initiatives. Even though China's pig industry is battling winter surges in disease (including the recent African swine fever outbreak), Agricultural Minister Tang Renjian is projecting reverting back to 2017 hog numbers by 2021 and seeks to maintain stable future pig populations and stop pig farmers from killing sows when prices dip.[110]

Besides digital technologies, China is also going for biotechnologies as a solution for its farming and food security needs. A key economic policy meeting in December noted that China will build a "national food security industry belt" to link up all Chinese major grain areas with the State Council's annual rural policy blueprint "No. 1 document."[111] It emphasizes seed sector development to augment food security, pushing major scientific projects in breeding and the "industrial application of biological breeding" (a term in China that relates to genetically modified or GM crops and other biotech applications).[112] Li Xinhai, head of Biotechnology Research Institute with the Chinese Academy of Agricultural Sciences, noted that China is self-sufficient in wheat and rice seed cultivation, while China-developed corn seeds make up 91% of domestic market share.[113] But, core future technologies in seed breeding, core technical tools of gene editing are still dominated by the US and therefore China needed to augment its basic research and cultivation technologies to catch up.[114]

Jiao Shanwei, editor-in-chief of cngrain.com website focusing on grain news, indicated that the No. 1 policy also shows that even though, in terms of crop seeds, China is still behind the US (with more than 35% of the global seeds market share).[115] China's crop seeds market size was growing at a rate of approximately 3% before 2016 which was lower than

[109] Harper, Justin, *op. cit.*

[110] Patton, Dominique and Hallie Gu, *op. cit.*

[111] *Ibid.*

[112] *Ibid.*

[113] GT Staff Reporters, *op. cit.*

[114] *Ibid.*

[115] *Ibid.*

the global average growth of 8.88% from 2005 to 2018.[116] Technologically, Chinese experts like Li Xinhai (Head of Biotechnology Research Institute with the Chinese Academy of Agricultural Sciences) argued that they are behind advanced nations in basic research in studying efficient breeding traits and expanding diversity of seeds for optimal seed security.[117]

Having detailed China's agricultural modernization and technological use, the following chapter adopts a comparative perspective to examine the agricultural sub-sector of eco-feed by comparing its development with those of neighbouring locations of Japan and Taiwan.

Bibliography

CGTN, "China moves to rural vitalization with focus on grassroots governance" dated 26 April 2021 in CGTN (downloaded on 26 April 2021). Available at https://news.cgtn.com/news/2021-04-25/President-Xi-Jinping-inspects-south-China-s-Guangxi-ZKMVrKjyeI/index.html.

CGTN, "How can China improve its food security?" dated 7 March 2021 in CGTN (downloaded on 7 March 2021). Available at https://news.cgtn.com/news/2021-03-07/How-can-China-improve-its-food-security--YnGmalQX1C/index.html.

Chen, Fu and John Davis, "Land reform in rural China since the mid-1980s" undated in Food and Agriculture Organization (FAO) (downloaded on 1 May 2021). Available at http://www.fao.org/3/x1372t/x1372t10.htm.

China Daily, "95% self-sufficiency urged for grains" dated 16 December 2013 in China Daily (downloaded on 1 January 2021). Available at https://www.chinadaily.com.cn/business/2013-12/26/content_17197186.htm.

Du, Jane, "China's Food Self-Sufficiency and Food Trade Dependency" dated 2 August 2018 in EAI Background Brief No. 1375 (Singapore: NUS East Asian Institute), 2018, p. 3.

GT Staff Reporters, "China to accelerate reform to agriculture, ensuring food security" dated 23 February 2021 in Global Times (downloaded on 23 February 2021). Available at https://www.globaltimes.cn/page/202102/1216300.shtml.

Gu, Hallie and Dominique Patton, "On the autofarm: China turns to driverless tractors, combines to overhaul agriculture" dated 16 January 2019 in Reuters (downloaded on 16 January 2019). Available at https://www.reuters.com/article/us-china-farming-technology-idUSKCN1PA0DV.

---

[116] *Ibid.*
[117] *Ibid.*

Harper, Justin, "Huawei turns to pig farming as smartphone sales fall" dated 19 February 2021 in BBC News (downloaded on 19 February 2021). Available at https://www.bbc.com/news/business-56121470.

Johnson, Ian, "Barred From Owning Land, Rural Chinese Miss Spoils of Country's Success" dated 26 September 2019 in The New York Times (downloaded on 26 September 2019). Available at https://www.nytimes.com/2019/09/26/world/asia/china-land-rights-farming.html.

Lin, Chia-nan, "Farmers using land for agriculture to receive subsidies" dated 10 January 2018 in Taipei Times. Available at https://www.taipeitimes.com/News/taiwan/archives/2018/01/10/2003685522.

Little, Daniel, "The High-level Equilibrium Trap" undated in University of Michigan website (downloaded on 1 January 2021). Available at http://www-personal.umd.umich.edu/~delittle/elvin.pdf.

Low, Nelson, "China's Appetite for Meat is Still Growing" dated 16 November 2020 in Reuters (downloaded on 16 November 2020). Available at https://www.reuters.com/article/sponsored/china-appetite-still-growing.

McCrimmon, Ryan, "China is buying up American farms. Washington wants to crack down" dated 19 July 2021 (downloaded on 19 July 2021). Available at https://www.politico.com/news/2021/07/19/china-buying-us-farms-foreign-purchase-499893.

McMillan, Tracie, "How China Plans to Feed 1.4 Billion Growing Appetites" dated February 2018 in National Geographic (downloaded on 1 January 2021). Available at https://www.nationalgeographic.com/magazine/article/feeding-china-growing-appetite-food-industry-agriculture.

Patton, Dominique and Hallie Gu, "China steps up focus on food security in major policy document" dated 22 February 2021 in Reuters (downloaded on 22 February 2021). Available at https://www.reuters.com/article/china-agriculture-document-idINKBN2AM0H5.

People's Daily Online, "Chinese farmers take advantage of high tech to increase production" dated 17 August 2020 in People's Daily (downloaded on 17 August 2020). Available at http://en.people.cn/n3/2020/0817/c98649-9732362.html.

Qin, Hengde, "Smart farming technology can transform Chinese agriculture and help feed the planet" dated 18 February 2021 in Global Times (downloaded on 18 February 2021). Available at https://www.globaltimes.cn/page/2021 02/1215818.shtml.

Sasada, Hironori, "Challenges in boosting Japan's agricultural exports" dated 12 March 2021 in East Asia Forum (downloaded on 12 March 2021). Available at https://www.eastasiaforum.org/2021/03/12/challenges-in-boosting-japans-agricultural-exports/.

Wong, John and Chen Gang, "Climate Change Posing a Threat to China's Long-Term Food Security" dated 13 August 2014 in EAI Background Brief No. 945 (Singapore: NUS East Asian Institute), 2014.

Xinhua, "Xi Focus: Xi charts road map for rural vitalization after victory in poverty fight" dated 25 February 2021 in Xinhuanet (downloaded on 25 February 2021). Available at http://www.xinhuanet.com/english/2021-02/25/c_139766552.htm.

Xinhua News Agency, "Outline of the People's Republic of China 14th Five-Year Plan for National Economic and Social Development and Long-Range Objectives for 2035" dated 12 March 2021 in Georgetown University Center for Security and Emerging Technology (downloaded on 12 March 2021). Available at https://cset.georgetown.edu/wp-content/uploads/t0284_14th_Five_Year_Plan_EN.pdf.

Zhou, Zheng, "China needs to diversify agricultural imports, provide investors with opportunity" dated 20 October 2019 in the Global Times (downloaded on 20 October 2019). Available at https://www.globaltimes.cn/content/1167374.shtml.

Chapter 5

Comparative Study of Eco-feeds: A New Type of Small-Sized Poultry Farming in Japan (Eco-feeding at Yokomine's Farm in Osaka) with Comparative References to China

In this chapter, China's development in high-value-added eco-feed for rearing chickens is compared with Japan's case study. It is a comparative study of how far China has developed its technologies in this sector. Some references are also made to Greater China developments like Taiwan, whose consumers like Japan, and increasingly China, are demanding higher quality premium organic foods for their consumption.

Introduction

Japanese consumers love eggs, with the average number of annual egg consumption per person being 366 in Japan. This is the second largest in the world (Mexico is in first place[1]). Japan is notorious for its expensive food prices; however, egg prices are an exception. The Japanese poultry industry succeeded in supplying eggs at lower prices by reducing labour costs through the introduction of automation technologies and reducing

---

[1]For more details, see Takagi, Shin-ichi, *Tamago Daijiten* (The Dictionary for the Egg Industry) (2nd version), Tokyo: Kogakusha, 2020.

feed costs by importing less expensive grains from foreign countries. Without a doubt, egg prices in Japan have had a downward trend since the post-Pacific War period. "Eggs are number one for cheapness" is a popular saying among Japanese consumers.

Comparative case studies in the current situation

In 2022, however, the majority of poultry farms in Japan suffered from the surge of oil and feed grain prices in the international market. Japan heavily relies on oil for electric power generation. Thus, with the increase in oil prices, electric power companies' charges increased to users, including farms. In recent years, there have been three ominous signs of growing concern for Japanese poultry farms. That is, increasing frequency of occurrence of communicable diseases, increasing criticisms on the raising style of Japanese poultry farms from the international animal welfare groups and increasing complaints from neighbours regarding the foul smell of the fowl droppings.

While egg consumption is on the rise for the very important domestic market of Japan, neighbouring China is exporting eco-feed to major developing economies like Egypt. Chinese chicken eco-feed ventures like New Hope company is the pioneering feed factory started in 2010 in Sadat City/Menoufia province/northern Cairo in Egypt by New Hope Liuhe Co. Ltd. (a ranking Chinese company specializing in feed, farming and food supplies) and it produces eco-feed for Egypt, as Yang Luofan (manager of quality control department) articulated the following:

"[New Hope imported feed from China increases broiler capacities to absorb nutrients from the feed and lessen nitrogen production]. We call the product biological feed, which is characterized by high-quality, eco-friendliness and efficiency."[2]

Background to the Japanese case study of Yokomine farm

However, not all success stories need to be large-scale like New Hope. There are Japanese farmers at the forefront of eco-friendly egg

---

[2]*Xinhua*, "Chinese company brings 'new hope' to Egypt's animal farming" dated 29 May 2022 in China Daily (downloaded on 29 May 2022). Available at https://global.chinadaily.com.cn/a/202205/29/WS6292b397a310fd2b29e5f90b_2.html.

production. Testuya Yokomine (aged 38 years) presents an exceptional case by establishing a unique style of poultry farming at Miyama, a small mountainous community in Osaka Prefecture. Yokomine prepares eco-feed from industrial wastes, which are collected without material cost; therefore, oil and foreign grain prices have no effect on Yokomine's feed cost. He is also not concerned about having animal welfare issues as well as the risk of communicable diseases because his poultry-raising style is healthy for the hens. The droppings from the hens at Yokomine's farm do not have an offensive odour and are used as materials for improving soil fertility in vegetable farms in the neighbourhood. How did Yokomine create such a unique raising style? This study presents a potential new type of poultry farming in Japan by studying Yokomine's case.

Yokomine's Experience at Iwaijima, a Small Isolated Island

Yokomine was born in 1984 in Yamaguchi Prefecture, the west end of the main island of Japan. Yokomine loved to go to an isolated place called Iwaijima as a small child, where his relatives lived. The islanders' lifestyle was in harmony with environment and nature. Yokomine enjoyed various field activities there. This experience strengthened his power in observing the habits of animals, which later became his advantage in raising hens.

In Iwaijima, a unique farmer, named Ujimoto, reared pigs for meat. Ujimoto's strategy was farming with minimal costs based on the honesty and kindness of the islanders. Obtaining all the islanders' permission, Ujimoto allowed his pigs to roam freely (with no fence or pen). Ujimoto collected leftover food from the islanders and fed it to his pigs. Yokomine was so impressed with Ujimoto's farming that he dreamed of doing a similar type of farming somewhere in Japan in future.

After Yokomine graduated from a technical school in 2006, he started working at an agribusiness company, called My Farm. The main business of My Farm was the operation of hobby farms for urbanites. Yokomine worked as an instructor for hobby farming for five years. Meanwhile, he found Miyama a favourable place for farming. Leaving My Farm in 2011, Yokomine moved to Miyama to start his own farm.

Miyama is about 40 km from the business district of Osaka City. In spite of its closeness to the highly urbanized area, Miyama is not urbanized because of its mountainous landscape (the elevation of Miyama is around 300 m).

As mentioned, Yokomine wanted to adopt Ujimoto's style of farming; however, this style of farming (i.e. allowing pigs to roam freely) was, and still is, only available in the small isolated community of Iwaijima. Considering the natural and social environment of Miyama, Yokomine decided to raise hens (instead of pigs). It should be noted that Yokomine's method for raising hens is different from any other poultry farm in Japan.

Japanese-Style Poultry Farming

We will discuss how hens are raised at a majority of Japanese poultry farms before talking more about Yokomine's farming methods. Table 1 shows that the average size of poultry farms has kept increasing. As of 2021, the average number of hens in a poultry farm is nearly 100,000 (Table 1). Poultry farms that hold more than 100,000 hens share nearly 75% of Japan's total egg production (Table 2). Most poultry farms in Japan (as well as large-scale production in many other countries) are so large and systematically controlled that they are often called "factories" (instead of farms). Henhouses are windowless and the length of daytime (and night-time) is created by electric lights. Each hen is kept in a small cage and is unable to move about freely. Unattended operations are dominant in most poultry farms with the introduction of automatic (and unmanned) feeding and egg collection machineries. The main ingredient of feed is foreign grains. A system such as this is called the Japanese-style poultry farming.

The Japanese-style poultry farming has two advantages. First, it is effective in reducing the production costs for eggs. This enables poultry farms to survive even with relatively lower prices of eggs. As seen in Figure 1, the egg prices have been on a downward trend since the 1950s. Second, Japanese-style poultry farming enables farmers to have stable production of eggs (e.g. number of eggs, colour, taste and weight of the whole egg). While livestock products are generally expensive in Japan, egg is exception (Table 3). Thus, the possibility of eggs being unavailable at major supermarkets and restaurants is minimized. In addition, the quality of eggs is so consistent that consumers can expect the same taste in all eggs purchased.

Table 1.   Total number of poultry farms and hens from 1960 to 2021

Year	(1) Total number of poultry farms	(2) Total number of hens (in thousand)	(3)=(2)/(1) (in thousand)
1960	3,838,600	54,627	0.014
1961	3,831,300	71,891	0.019
1962	3,805,600	90,006	0.024
1963	3,722,500	98,925	0.027
1964	3,496,000	107,738	0.031
1965	3,243,000	120,197	0.037
1966	2,767,000	114,500	0.041
1967	2,508,000	126,043	0.050
1968	2,192,000	140,069	0.064
1969	1,941,000	157,292	0.081
1970	1,703,000	169,789	0.10
1971	1,373,000	172,226	0.13
1972	1,058,000	164,034	0.16
1973	846,400	163,512	0.19
1974	660,700	160,501	0.24
1975	509,800	154,504	0.30
1976	386,100	156,534	0.41
1977	328,700	160,550	0.49
1978	278,600	165,675	0.59
1979	248,300	166,222	0.67
1980	n.a.	n.a.	n.a.
1981	187,600	164,716	0.88
1982	160,600	168,543	1.0
1983	145,300	172,571	1.2
1984	134,300	176,581	1.3
1985	124,100	177,477	1.4
1986	117,100	180,947	1.5
1987	109,900	187,911	1.7
1988	103,000	190,402	1.8
1989	95,200	190,616	2.0
1990	87,200	187,412	2.1

(*Continued*)

Table 1.   (*Continued*)

Year	(1) Total number of poultry farms	(2) Total number of hens (in thousand)	(3)=(2)/(1) (in thousand)
1991	10,700	188,786	18
1992	9,770	197,639	20
1993	9,070	198,443	22
1994	8,420	196,371	23
1995	7,860	193,854	25
1996	7,310	190,634	26
1997	7,020	193,037	27
1998	5,840	191,363	33
1999	5,520	188,892	34
2000	5,330	187,382	35
2001	5,150	186,202	36
2002	4,760	181,746	38
2003	4,530	180,213	40
2004	4,280	178,755	42
2005	n.a.	n.a.	n.a.
2006	3,740	180,697	48
2007	3,610	186,583	52
2008	3,430	184,773	54
2009	3,220	180,994	56
2010	n.a.	n.a.	n.a.
2011	3,010	178,546	59
2012	2,890	177,607	61
2013	2,730	174,784	64
2014	2,640	174,806	66
2015	n.a.	n.a.	n.a.
2016	2,530	175,733	69
2017	2,440	178,900	73
2018	2,280	184,350	81
2019	2,190	184,917	84
2020	n.a.	n.a.	n.a.
2021	1,960	183,373	94

Source: Ministry of Agriculture, Forestry and Fisheries.

Table 2.   Number of poultry farms and hens as of 2018

	National total	Farm size measured by the numer of hens				
		From 1,000 to 4,999	From 5,000 to 9,999	From 10,000 to 49,999	From 50,000 to 99,999	100,000 or more
Number of farms	2,200	536 (26.9)	287 (14.4)	613 (30.7)	226 (11.3)	332 (16.6)
Number of hens (in thousand)	139,036	1,322 (1.0)	2,048 (1.5)	15,264 (11.0)	15,832 (11.4)	104,515 (75.2)

Notes: Within parentheses is the perecentage related to all the farms.
The national total includes research institutes that were not measured by farm size.
Source: Ministry of Agriculture , Forestry and Fishries.

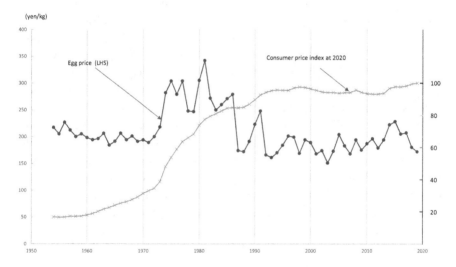

Figure 1.   Egg price in comparison with consumer price index
Source: Takaki [2020], Ministry of Internal Affairs and Communications.

Conversely, the Japanese-style poultry farming also has disadvantages. The biggest problem is that hens are confined in unhealthy conditions. A cage is too small to have physical exercise. Imported grains, which share a large portion of feed, contain chemicals for preservation, which is detrimental to the hens' health. As a result, hens are prone to communicable diseases. Furthermore, eggs laid by unhealthy hens do not

Table 3.   Comparison of major livestock products between cities as of November 2006

		Yen	Converted price in Japanese yen		
Name of product	Unit	Tokyo	New York	London	Paris
Beef (roast)	100g	386	295	332	279
Pork (shoulder)	100g	159	144	112	142
Egg (large size)	10 pieces	222	310	309	418
Milk	I litre	182	173	142	175

Notes: Exchange rates are: 1 USD = 118.41 yen, 1 GBP = 228.28 yen, and 1 euro = 152.63 yen.
Source: Economic Planning Agency.

taste well. Such challenges apply to other large-scale captive poultry farms in other countries as well.

Treatment for fowl droppings is also a big problem in the Japanese-style poultry farming. Fowl droppings from healthy hens can be processed into manure fertilizer, which can be sold to vegetable farms for improving soil fertility. However, hens in Japanese-style poultry farming are oftentimes so unhealthy that fowl droppings should be treated as useless waste, which should be detoxified (a cost factor) before disposing. Fowl droppings from unhealthy hens exude a disagreeable smell. This generates serious conflicts between poultry farms and their neighbours.

Currently, unlike the European and North American countries, animal welfare regulations on poultry farming are weak in Japan and animal welfare activists criticize Japanese-style poultry farming. There is a possibility that the animal welfare issue may be a growing issue for the Japanese poultry industry.[3] Meanwhile, in contrast, the priority in large-scale Chinese eco-feed conglomerates is to feed large populations in the developing world. With regard to the New Hope case study, Ibrahim Hassan, a 36-year-old driver who was interviewed by *Xinhua*, revealed: "The number of poultry birds in our farm has reached 3 million, two-thirds of which are fed by feed produced by New Hope."[4] Chinese eco-feed criteria focus on feed conversion ratio as a gauge of poultry or livestock production

---

[3]In 2018, an executive officer of the Japan Poultry Association (Japan's biggest association among poultry farmers) secretly donated money to Takamori Yoshikawa, the then-Minister of Agriculture, Forestry and Fisheries, to prevent the Ministry from introducing strong animal welfare regulations to the industry. In 2020, this donation was uncovered and Yoshikawa was accused of receiving an illegal donation. This scandal reflects the fact that the pressure to advance animal welfare is mounting, even in Japan. (https://www.asahi.com/ajw/articles/14049187).

[4]*Xinhua, op. cit.*

efficiency (i.e. ratio of given feed weight over animal weight gain in a certain period of time) as Su Hao (General Manager of New Hope Egypt Co. Ltd.) explained: "The lower the feed-to-meat ratio, the higher the level of production efficiency and economic benefits. Before we came here, the ratio in the Egyptian feed industry was 1.8, but now, using our feed, the figure has dropped to 1.5."[5]

Another priority of large-scale Chinese eco-feed producers like New Hope is to generate jobs for the local economy, e.g. Mohamed Ramadan was recruited into New Hope Egypt Co. Ltd. in 2012, and he is now deputy manager of the production department, he said: "When I first joined New Hope, I was a warehouse manager. Through receiving training and continuous learning at the company, I gradually mastered the advanced feed production technology and eventually became a member of the company's management in Egypt."[6]

Yokomine's Unique Style of Raising Hens

Yokomine's method of raising hens is entirely different from the Japanese-style poultry farming. Yokomine's basic idea is to keep hens healthy by providing sufficient space and high-quality feed. As such, he employs no cages and provides plenty of walking space for hens. Yokomine's hen-houses are simple with plenty of windows that allow sunlight into the hen-houses. Yokomine believes this helps hens acquire resistance to diseases.

In addition, Yokomine prepares high-quality eco-feed on his own. Since Miyama is close to the business district of Osaka, Yokomine finds various types of industrial wastes within his driving range daily, which can be processed into feed. For example, he uses strained lees from beer breweries and cooking oil factories, skulls and bones from restaurants and broken charcoal.

Producers of these industrial wastes have the responsibility to dispose them without damaging the natural environment. Thus, it is favourable for them if Yokomine disposes these wastes for them through his use. As a result, he is allowed to collect these wastes without any charge as much as he would like. The volume and quality of these wastes are unpredictable and differ day by day. Thus, processing these industrial wastes to feed requires special skills. Fortunately, Yokomine has rich knowledge of food science.

---

[5] *Ibid.*
[6] *Ibid.*

Consequently, Yokomine's hens are quite healthy and the eggs from his farm are so delicious that the eggs price at nearly 100 yen per piece, while ordinary eggs are sold from 10 to 20 yen per piece, at supermarkets.

Fowl droppings at Yokomine Farm do not exude a bad smell and can be used as materials for improving soil fertility at vegetable farms. Yokomine increases his popularity in Miyama by offering his fowl droppings to vegetable farms in his neighbourhood.

Receiving good care from Yokomine, hens on the farm live (and lay eggs) long. While hens in Japanese-style poultry farming live only three years, hens in Yokomine's Farm live nearly six years under healthy conditions.

The total acreage of Yokomine's henhouses is around 200 a (= 0.2 ha). Yokomine raises nearly 300 hens there. While the number of hens is substantially smaller than the Japanese-style poultry farming, Yokomine Farm generates sufficient profits by gaining high prices and saving costs for feed.

Chinese Case Studies of Chicken Eco-Feed: Perhaps, China's equivalent of the Yokomine farm is Hu Jinyan [retired local middle school instructor and General Manager (GM) of her breeding base farm in northern China Tianjin Municipality, 58 years old] utilize bran and corn to feed her chickens in addition to herbs and worms: "Our chickens run freely in the mountain. They eat worms, potherbs and wild fruits during the day, and we feed them mixed corns and bran without any additives at night."[7] Like Yokomine, the health of the poultry was paramount. Hu kept more than 50,000 white bellies, black tails and red cockscombs on the 200-hectare farm made up of approximately 30 chicken sheds constructed on mountaintop and, together with the eco-feed, she believed that food safety is ensured for young consumers' health, given that chickens and eggs are connected with daily living.[8]

Because of the eco-feed and mountaintop lifestyle, the chickens have emerged healthy for customers who are keen on organic foods. Hu articulated the following: "These lively chickens are good at running and flying compared with others and have a greater resistance to diseases. More importantly, the mountain land is a wonderful place to raise them.... In the future, I plan to develop part of the breeding base as an ecological tour,

---

[7]*Xinhua*, "Across China: High-tech helps ecological chicken-raising lay golden eggs in green mountains" dated 17 August 2019 in Xinhua (downloaded on 17 August 2019). Available at http://www.xinhuanet.com/english/2019-08/17/c_138316354.htm.
[8]*Ibid.*

combining poultry farming with tourism. My ultimate goal is to increase the value of this region's clear waters and green mountains"[9] Customer Wang Zhenxing (aged 33, Tianjin villager, Hu's loyal customer) purchased 1000 chickens from Hu in February 2019 after coming across a TV documentary about Hu's farm:

> Organic food is in high demand in the market and incorporating high-tech into raising chickens is promising [adding Hu has provided support to Wang and 96% of Hu's chickens are healthy].[10]

In addition to Yokomine's farm where eco-feed is used, Hu Jinyan's farm employs Industry 4.0 technologies. Hu utilizes smartphone real-time vid-feeds of the farm's conditions, egg-laying monitoring bracelets, WeChat (the App X of China) product tracking account.[11] Perhaps, another feature present in Hu's case was the help from the local government in the context of a state-centred approach to farming in China. The local authorities constructed roads in mountains and provided water and electricity along with expert coaching, as Hu revealed: "When it rained, the wheels of egg-carrying carts got stuck into the mud. It was hard to transport the eggs [an issue that was fixed]."

Greater China Comparative Study: Taiwanese Agriculture in an Organic World: Both features of organic food movement as well as use of high-tech farming implemented in Hu's farm can also be found in the Greater China agricultural context as well. It may be useful to examine another ethnic Chinese society's development of organic food/agricultural technologies. The organic movement has reached Taiwanese consumers. Enterprising Taiwanese organic retail chain companies like Cotton Land are a mature retail platform with well-stocked imported produce to serve their customers while Taiwanese consumers are seeking tasty cuisines that are equally healthy and high quality in a land well-known for its foodie fads.[12] In addition to the foodie experience, some are interested in produc-

---

[9] *Ibid.*

[10] *Ibid.*

[11] *Ibid.*

[12] Chang, Chiung-fang, "Farming for a Toxin-Free World" dated March 2014 in Taiwan Panorama (downloaded on 1 January 2020). Available at https://www.taiwan-panorama. com/ja/Articles/Details?Guid=28403497-8b48-452c-a0c6-af9c6ded02ab&langId= 3&CatId=9.

ing fresh foods themselves. For entrepreneurial-minded individuals, Taipei's organic crops-related tourist farms provide the DIY (Do It Yourself) organic farming and fruit harvesting programmes, e.g. the iconic Baishihu (白石湖) tourist farm in the Neihu District located in the peripheries of Taipei.[13] In addition, it appears age is of no consequence to starting DIY organic farming ventures. Cho Chen-ming (卓陳明), elementary school educator and garment entrepreneur turned owner of Toucheng Leisure Farm (頭城休閒農場) in her 80s, retired to the rural areas at 40 years old to create an organic farm that is also a site of nostalgia revival, and today, this humble facility hosts 60,000 global visitors annually with a professional Swiss chef retiring there.[14]

Besides individual consumers and entrepreneurs, there is institutional clientele driving the consumption of organic agricultural produce as well. Major Taiwanese chain distributors of organic products include retailers, super-mart chains, warehouse distributors and even cultural establishments like Leezen, Cotton Land, Pxmart, Wellcome, Matsusei, Carrefour, RT-Mart, a.mart, Eslite Books' Songshan Cultural and Creative Park branch.[15] Other than cultural facilities, organic food can also be found in eco-touristy cafes like the spacious Farmhouse Cafe TRATTORIA (穠舍田園咖啡餐廳), a gentrified vintage barn with a Mediterranean feel that looks down from the Neihu hills to the Taipei Basin, serving strawberry muffins and au lait produced with organic berries.[16]

Large super-mart chains acquire organically certified and traceable (defined as having no minute traces of 251 agricultural chemicals) products contractually.[17] The economy of scale brought about by their purchases depresses the prices of organic products down (e.g. Pxmart retails organic vegetables at affordable prices, a feature of its corporate branding)

---

[13] Travel Taipei Magazine, "Taipei's Organic Agriculture Mountain Tourist – Farm Experience in the Baishihu Area" dated 13 December 2020 in Taiwan Scene Taiwan Digital Travel Magazine (downloaded on 13 December 2020). Available at https://taiwan-scene.com/taipeis-organic-agriculture-mountain-tourist-farm-experience-in-the-baishihu-area/.

[14] Chen, Yi-han (translated by Jack C. and edited by Sharon Tseng), "Yilan Organic Farm Built by 80-year-old Grandma, Popular Among International Tourists" dated 30 October 2019 in Commonwealth Magazine Group (CMG) (downloaded on 30 October 2019). Available at https://english.cw.com.tw/article/article.action?id=2581.

[15] Chang, Chiung-fang, *op. cit.*

[16] Travel Taipei Magazine, *op. cit.*

[17] Chang, Chiung-fang, *op. cit.*

and these contracts give organic farmers an ease of mind in tending to their crops without being anxious about sales.[18] Conventional crops that have a higher chance of having pesticide residues like beans, peas, bell peppers, cucumbers, Arden lettuce and baby bok choy represent an opportunity for organic farmers as supermarket chains like Pxmart acquire such products from contracted traceable and certified farmers.[19] Cultivating organic crops that are more delicate than those enhanced with chemicals also requires special facilities like greenhouses to house them but capital is needed to set up these controlled environments.

Against the backdrop of agricultural economies jostling for a greater market share to capitalize on a global trend in living a sustainable lifestyle, the Taiwanese COA assists organic farmers in building greenhouses and applying for organic certificates.[20] Besides the private sector, even the government is in the act of helping the industry grow. In 2019, the Taiwanese authorities also called for more consumption of organic agricultural products in schools, armed forces and public sector departments, further coordinated by a unit under the COA that takes charge of initiatives like disseminating organic education, training of specialists and marketing help for organic farmers.[21]

Taiwanese consumers want to ensure that food safety is given high priority and that the food they eat is free from toxic chemicals, and some like Yang Rumen, founder of Taipei's 248 Farmers' Market, opined the following:

> Pesticide-free and organic produce sold very well [in 2013].... The market provides producers and consumers with a communications channel.... But some of the older farmers find talking to people far more difficult than growing crops.... Take the pricing myth. Everybody thinks that 'organic' means 'expensive,' and therefore only the wealthy can afford organic produce.... The fact is that people throw away one-third of the food they purchase. Why not spend a little more money to buy better

---

[18] *Ibid.*

[19] *Ibid.*

[20] Lin, Chia-nan, "Organic farming promotion act takes effect" dated 30 May 2019 in Taipei Times (downloaded on 30 May 2019). Available at https://www.taipeitimes.com/News/taiwan/archives/2019/05/30/2003716039.

[21] Huang, Tzu-ti, "Taiwan to boost organic farming industry with new law taking effect May 30" dated 31 May 2019 in Taiwan News (downloaded on 31 May 2019). Available at https://www.taiwannews.com.tw/en/news/3714796.

quality food, and then eat it all?.... Our hope is that by communicating with the public in farmers' markets, farmers can, over the course of a year or two, develop a reliable sales channel.[22]

Food safety concerns have even triggered some to become entrepreneurs to take charge of their own lives. Martin Lin (林清立), a global trading firm boss turned organic farm owner, was inspired to make the career switch due to health and food safety issues that motivated him to return to his native Baishihu in the early 2000s to work on his inherited long-unused farmland:

> In light of my own health issues, I had reservations about the produce I was consuming, how I was hurting my own land, and how I might be injuring my own customers.... Only if you dig your fingers into the soil do you develop intimacy with the earth — and a protective love that will last the rest of your life.[23]

In the mid-2010s, a number of food safety-related media expose caused the utilization and sales of "Earth friendly pesticide-free agricultural produce" to increase dramatically.[24] Thus, farmers' market like 248 imposes the conditionality of no chemical pesticides or fertilizers for crops sold in their facility, something that attracts hotels, restaurants and organic shops to their stalls with some retailers even becoming household names.[25]

Taiwan's Organic Agriculture Promotion Act (有機農業促進法) effective from Thursday 30 May 2019 governs culpability for crop pollution and contamination whereby, under the law, pollution and contamination originating from neighbouring farmlands would not result in punitive measures though the polluted/contaminated crops cannot be retailed anymore as "organic" produce.[26] With growing awareness of climate change, trendy desire for detoxified lifestyles and food safety, demand for zero-pesticide agriculture among socially conscious, direct-sales organic farmers markets and warehousing distributors offering such products have become popular, allowing some farmers to tend to their crops in their

---

[22]Chang, Chiung-fang, *op. cit.*

[23]Travel Taipei Magazine, *op. cit.*

[24]Chang, Chiung-fang, *op. cit.*

[25]*Ibid.*

[26]Lin, Chia-nan, *op. cit.*

farms as consumers seek them out for supplies.[27] The trend has even extended to an organic farming eco-tourism boom. There are already areas in Taiwan that specialize in organic tourist farm experiences. Over the past two decades, Baishihu has become a premium sustainable organic farming region from 2010 to 2020 with iconic veterans like Martin Farm (野草花果有機農場) located in the elevation up Bishan Road from the flatlands that is widely acknowledged to have started the organic farm boom.[28]

The organic food movement

As a liberal democracy, Taiwanese farmers also enjoy tremendous freedom in assembling together like-minded small farmers in grassroots initiatives for organic agricultural activities. Fruit cultivator turned Yilan rice and chicken farmer Lee Pao-lien (nicknamed "Ah-pao") organized a circle of "Earth-friendly" small farmers, explaining:

> Farmers' markets provide small farmers with a platform, a forum in which they can explain themselves to the public … [and she's focusing her energies on convening the Protect Yilan Workshop, a group working on environmental issues pertaining to farming, such as field ecology and pollution from agricultural runoff].[29]

Similarly, chicken farmer Chen Weiren from Gaoshu Township Pingtung County eliminates antibiotics, growth hormones and preventive medications for breeding "delicious, toxin-free" poultry.[30]

It appears Taiwanese organic farmers have very personal and varied motivations for going into organic farming. Some are more entrepreneurial in nature than others. Cho Chen-ming (卓陳明), elementary school educator and garment entrepreneur turned owner of Toucheng Leisure Farm (頭城休閒農場) in her 80s explained the conceptual origins of her organic farm:

> "People often ask me: how do you determine what's 'organic' food? Actually, all you need to do is observe if the soil is organic and fresh. Grab a handful of soil and see if it is loose and fluffy, then you will know! If the

---

[27] Chang, Chiung-fang, *op. cit.*
[28] Travel Taipei Magazine, *op. cit.*
[29] Chang, Chiung-fang, *op. cit.*
[30] *Ibid.*

soil is tough, that means there are no worms, microorganisms, or anything living inside. You cannot grow organic food in such soil.... I really like the farm life. In the beginning, I bought this place for myself and my family. It just so happened the government was promoting a policy that supported leisure farms. We followed their footstep, but we never exploited the mountain.... In truth, I was conned into buying this piece of the mountain!... I don't have a lot of money, but I want a big piece of land.... The place was a wasteland back then, with grass as tall as people. The agent pointed into the distance and told me, all that land was mine. Only after we signed the deal did I discover — it didn't include the piece of flat land we were standing on! I had to pay extra to get the complete plot."[31]

Taiwanese organic farmers are relentless in promoting environmentally friendly toxin-free agriculture (defined as the absence of pesticides and chemical fertilizers) through farmers' markets, supermarkets, direct sales and the Internet while distribution outlets like 248 Farmers' Market have necessitated all their participating farmers to be present at the market weekly regardless of weather conditions as a responsibility and promise kept to customers.[32] The farmers' markets facilitate growers and consumers to interact directly with each other to strengthen customer confidence in directly buying the produce, paying a little higher prices for their food safety concerns.[33] They advocate for the zero-pesticide movement and boosting CSA in retailing local produce through personal contact rather than reliance on official certifications (e.g. Yilan's Ko Tong Rice Club and Hsinchu's Qianjia CSA).[34]

Chen Jian-tai (head of Qianjia CSA that comes under the Industrial Technology Research Institute's Social Welfare Committee serving the suburban Hsinchu Science Park university towns and high-tech workers with 20 produce, such as beans, tomatoes and tubers) articulated:

I used to work in the Internet sector.... For all that I now run a farm, what I'm really doing is moving rice sales online.... On average, we cultivate our crops for a week more than conventional growers.... With time to mature, they have more flavour.... The 'Earth-friendly' movement is very powerful."[35]

---

[31] Chen, Yi-han (translated by Jack C. and edited by Sharon Tseng), *op. cit.*
[32] Chang, Chiung-fang, *op. cit.*
[33] *Ibid.*
[34] *Ibid.*
[35] *Ibid.*

Finally, these community initiatives were given the legal backing of the law. Taiwan's Organic Agriculture Promotion Act (有機農業促進法) came into being on Thursday 30 May 2019 to protect water and soil, the environment, biodiversity, animal welfare, consumer interests and legally promote eco-friendly and sustainable use of resources.[36]

"Scholar-farmers" have utilized their writing skills to advocate organic agriculture establishing Ko Tong Rice Club in Yuanshan Township Yilan County in 2004.[37] Yang Wen–Chuan is an example of a "scholar farmer" who was formerly from National Taiwan University doing rural village planning before working on 1.5 hectares of leased land in Yuanshan Neicheng Village as the founder of "200 Hectares" agricultural incubator that secures farmlands and dispenses advisories to new farmers.[38] From professors to graduate students, these scholar farmers bring new blood, fresh ideas and technological experts to the agricultural field. For individuals like Yang, the authorities put in place additional policy assistance through subsidies and material support. For example, the 2019 Minister of the COA Chen Chi-chung (陳吉仲) has in place subsidies between NT$30,000 and NT$80,000 ('NT$' stands for New Taiwan Dollar) per hectare for cultivators transitioning to eco-farming, with machinery and organic fertilizer support and 40% discount in rental fees for utilizing public lands for organic agricultural activities.[39]

There are other examples of scholar farmers from the graduate student community. Wu Jialing who hails from a farming family in Yunlin stopped her postgraduate studies at Shih Hsin University Graduate Institute for Social Transformation Studies to turn to professional rice farming on four hectares of leased land.[40] She founded its "Fields and Rice Workshop" brand (managing farm work like germination, transplantation, weeding, transplantation and snail removal until the July harvest, followed by marketing activities thereafter): "I want to change the public's image of farming.... Growing rice isn't hard.... It just takes three phone calls."[41]

---

[36] Huang, Tzu-ti, *op. cit.*
[37] Chang, Chiung-fang, *op. cit.*
[38] *Ibid.*
[39] Huang, Tzu-ti, *op. cit.*
[40] Chang, Chiung-fang, *op. cit.*
[41] *Ibid.*

Concluding remarks

By April 2019, 12,194 hectares (approximately 1.5% of Taiwan's cultivated lands) were utilized for organic or eco-friendly farming.[42] Taiwan has 6000 hectares of organic farmland (0.6% of overall arable land under cultivation), compared with France's 16% and Austria's 20%, thus there is still much potential to go for Taiwan's organic industry.[43] Competition with other agricultural entities is given a boost by the legislative system as laws were put in place to ensure equity. The Organic Agriculture Promotion Act (有機農業促進法 promulgated on 30 May 2018) also ensures reciprocity as it requires other parties selling organic produce to Taiwan to sign equivalence treaties within 365 days of 30 May 2019 to protect the rights of domestic organic farmers.[44] But area size alone may not be an important determinant factor for the contemporary organic farmer. Today, Taiwanese organic tenants have the world of technologies laid down at their feet. Rice cultivation, for example, is mechanized and comes with hired contractors who turn the earth, transplant seedlings and harvest the product via smartphone and then utilize the Internet to retail the rice, interact with consumers via Facebook updates and social media followers on the site.[45]

Conclusion

For years, the Japanese poultry industry has developed successfully by expanding farm size, introducing mechanization and relying on imported feed grains. However, this made the Japanese poultry industry vulnerable to price fluctuations in the international oil and grain markets. In addition, The Japanese-style poultry farming is increasingly criticized for not satisfying animal welfare conditions.

Yokomine Farm presents an antithesis to the Japanese-style poultry farming. Yokomine provides plenty of space for hens in henhouses and prepares high-quality eco-feed at a low cost. Yokomine's unique experience at Iwaijima helped him to establish a new system for raising hens. Developing a poultry ecosystem that can create and accommodate highly

---

[42] Lin, Chia-nan, *op. cit.*

[43] Chang, Chiung-fang, *op. cit.*

[44] Lin, Chia-nan, *op. cit.*

[45] Chang, Chiung-fang, *op. cit.*

talented farmers such as Yokomine may be a worthy challenge for today's Japanese poultry industry.

While Yokomine's farm started off as a hobbyist business and then as a model producer of eco-feed farming, Hu's farm has quickly professionalized and turned into a showcase model. Hu received professional advice from Professor Zang Sumin from Hebei Agriculture University, with Prof Zang articulating: "Hu was not professional at first; she always chose big-sized chickens with beautiful feathers. She didn't know that some small-sized chickens tended to have better abilities to lay eggs [who gave advice on breeding, epidemic prevention, environmental monitoring and management for Hu]."[46] The use of technology apparently increased productivity. The electronic bracelet monitors' scientific breeding approach increased the egg output rate from 30% to 60%, enabling Hu to trademark her base in 2012 in the form of a complete industry chain (poultry hatching, breeding and selling) and, in 2018, the total revenues attained approximately 11 million yuan (US$1.56 million, including 50% internet sales).[47] The tables then turn and she became a mentor and coach to other villager farmers without fees, inspiring followers from Hebei, east Shandong and Jiangsu provinces.[48]

The eco-feed industry has great potential for export (with economic diplomatic implications) as the case study of New Hope shows. Shao Wen (head of New Hope Egypt of the overseas business department of New Hope Liuhe Co. Ltd.) noted that New Hope has four feed-making sites and one Egyptian firm worth 600 million Chinese yuan (US$90 million) producing 800 positions for local Egyptians.[49] This set-up is a part of the BRI economic diplomatic policy as articulated by Shao:

> We closely follow the steps of the BRI, supported by more than 40 years of expertize of our Chinese technical teams and using leading production technologies to process high-quality feed, which has played a role in the development of Egypt's agriculture and animal husbandry industry.[50]

---

[46]*Xinhua, op. cit.*
[47]*Ibid.*
[48]*Ibid.*
[49]90% of the employees in the company are Egyptians. (*Source*: *Xinhua, op. cit.*).
[50]*Xinhua, op. cit.*

Appendix

Advantage of Miyama: Low Risk of Wild Animal Attacks on Hens

Recently, the risk of wild animal attacks on livestock animals (including hens) is increasing. This is attributable to Japan's past forest management practices. In the 1950s, when Japan suffered a shortage of building materials, the Japanese government promoted planting cedar seedlings in forests by providing subsidies since it was suitable as construction material. This could be used as building materials for at least 50 years after planting the seedlings. However, those cultivators who planted cedar seedlings in the 1950s gave up thinning out cedar forests in the 1960s, when Japan liberalized the imports of cheap construction materials from foreign countries. As a result, many Japanese mountains are now covered with skinny (but mature) cedars, which are difficult to use in everyday life. This also disrupted the ecosystem in forests. Many wild animals have lost their habitats (and food sources) and wandered to the farms, preying on livestock animals.

Unlike other lumber areas during the fast-growth economic period, people in Miyama used the forests in a different way. It was known for producing a local speciality of *kanten* (Japanese gelatin made from *tengsa* seaweed) in winter. *Tengsa* seaweed is repeatedly boiled and air-cooled in a caldron, which is the traditional way of producing *kanten*. Since Miyama experiences tough cold winter winds, it was suitable for cooling *kanten*. Additionally, people in Miyama needed firewood for boiling *tengsa* seaweed. However, cedar is not suitable for firewood; therefore, the people of Miyama did not plant cedar seedlings. Consequently, they preserved the wild animals' habitats in the forest, reducing the risk of wild animal attacks on livestock animals in Miyama.

After examining China's manufacturing and agricultural firms/companies' use of technologies domestically, the final chapter in the section on China focuses on the external application of its technologies (and the potential for exportation) for other ASEAN/East Asian economies.

Bibliography

Chang, Chiung-fang, "Farming for a Toxin-Free World" dated March 2014 in Taiwan Panorama (downloaded on 1 January 2020). Available at https://www.taiwan-panorama.com/ja/Articles/Details?Guid=28403497-8b48-452c-a0c6-af9c6ded02ab&langId=3&CatId=9.

Chen, Yi-han (translated by Jack C. and edited by Sharon Tseng), "Yilan Organic Farm Built by 80-year-old Grandma, Popular Among International Tourists" dated 30 October 2019 in Commonwealth Magazine Group (CMG) (downloaded on 30 October 2019). Available at https://english.cw.com.tw/article/article.action?id=2581.

Huang, Tzu-ti, "Taiwan to boost organic farming industry with new law taking effect May 30" dated 31 May 2019 in Taiwan News (downloaded on 31 May 2019). Available at https://www.taiwannews.com.tw/en/news/3714796.

Lin, Chia-nan, "Organic farming promotion act takes effect" dated 30 May 2019 in Taipei Times (downloaded on 30 May 2019). Available at https://www.taipeitimes.com/News/taiwan/archives/2019/05/30/2003716039.

Travel Taipei Magazine, "Taipei's Organic Agriculture Mountain Tourist — Farm Experience in the Baishihu Area" dated 13 December 2020 in Taiwan Scene Taiwan Digital Travel Magazine (downloaded on 13 December 2020). Available at https://taiwan-scene.com/taipeis-organic-agriculture-mountain-tourist-farm-experience-in-the-baishihu-area/.

Shin-ichi, *Tamago Daijiten* (The Dictionary for the Egg Industry) (2nd version), Tokyo: Kogakusha, 2020.

Xinhua, "Across China: High-tech helps ecological chicken-raising lay golden eggs in green mountains" dated 17 August 2019 in Xinhua (downloaded on 17 August 2019). Available at http://www.xinhuanet.com/english/2019-08/17/c_138316354.htm.

Xinhua, "Chinese company brings 'new hope' to Egypt's animal farming" dated 29 May 2022 in China Daily (downloaded on 29 May 2022). Available at https://global.chinadaily.com.cn/a/202205/29/WS6292b397a310fd2b29e5f90b_2.html.

Part III

Exporting Technologies Outside China

Chapter 6

China's Capacity Building Roles in ASEAN's People-to-People Connectivity

The previous chapters mainly concentrated on China's domestic utilization of technology for development before discussing comparative perspectives in the preceding chapter. This chapter changes the focus to an external gaze and discusses China's tech interactions with its neighbouring region of Southeast Asia. With rapid technological developments, China and its state and private sectors are now ready to assist other countries with their own developments. Perhaps the region that stands to gain from this is the nearest region to China, its backyard of Southeast Asia. This chapter examines such roles in the context of ASEAN's desire for connectivity.

Theoretical ideas of ASEAN Connectivity: Central to the concept of people-to-people connectivity is the idea of common consciousness of shared heritage and culture. Conceptually, education, culture and tourism form the three legs of this idea of common consciousness among ASEAN people. People-to-people connectivity is the soft component of the hardware behind ASEAN connectivity. The "hard" aspects of ASEAN connectivity include physical infrastructures and institutional arrangements like free trade agreements. Therefore, people-to-people connectivity acts like a lubricant to smoothen out the cogwheels and mechanical machinery of ASEAN integration.

There are tangible and intangible elements behind the idea of people-to-people connectivity. Regardless of the categories they fall under, the basic principle behind people-to-people connectivity is enhancing people's mobility, mobility geographically and mobility across socioeconomic classes through education. The tangible features include quantifiable aspects of tourism and student arrivals. Educational aspects of people-to-people connectivity can be also quantified through people movements and educational revenues for educational institutions or even the number of entrants into the ASEAN workforce.

The intangible aspects include culture and historical/heritage conservation that cannot be easily quantified in some ways. They are dependent on memory retention, resonance with the audience, socialization processes and other social processes. Very often conservation of a particular heritage site involves work related to protecting both tangible and intangible assets. Protection of a natural landscape of a particular local community is often accompanied by preserving traditional aspects of the use of that land.

Operational Details: Within the ASEAN Masterplan for Connectivity, tourism is considered as "sectoral body" contributing to overall integration. It is a functional entity with very pragmatic purpose to serve in the overall integration plans. Therefore, operational details become significant and important. For example, to operationalize the idea of people-to-people contacts, relaxation of two weeks requirement for visa-free travel within ASEAN is conceptualized in the Master Plan on ASEAN Connectivity. Mutual Recognition Arrangements (MRAs) in the area of education are another incremental step towards operationalizing people-to-people contacts.

Various agreements are also in place for people mobility in terms of integrating transportation networks: Roadmap for Integration of Air Travel Sector (RIATS), and Roadmap Towards an Integrated and Competitive Maritime Transport in ASEAN (RICMT). Functionally, AFAFGIT is installed to "Establish an efficient, effective, integrated and harmonized transit transport system in ASEAN"; RIATS advocates, "Full liberalization in air transport services towards realizing Open Skies Policy in ASEAN"; RICMT, "Promote progressive liberalization of maritime transport service."[1] Besides policy changes to increase connectivity and

---

[1] ASEAN Secretariat, Masterplan on ASEAN Connectivity (Jakarta: ASEAN Secretariat), 2011, p. 18.

intra-ASEAN transport agreements, technology can also be harnessed to increase awareness of pan-ASEAN culture and heritage. Virtual resource learning centres established online are linked to each other to share more information about ASEAN's heritage sites and historical and cultural assets.[2]

Still on the technological point, it may be possible to continue with the idea of "APT Cyber University" conceptualized by the NEAT Working Group on the Enhancement of Cultural Exchange in East Asia in 2007–2012 with the intention to increase comprehension of different cultures and diversity.[3] Cyber university courses, virtual classrooms and other IT-based learning resources can encourage self-directed learning. It overcomes the need for funding and time to travel to developed economies to pick up knowledge and skills.

In the case of ASEAN countries like Singapore, its universities are also encouraging lifelong learning. University education therefore begins and not ends with graduation, therefore skills are continually upgraded throughout one's career and even into the retirement period. The state can assist with such constant upgrading by offering incentives and credits for adult learners to pick up new skills or refresh existing ones. In such lifelong learning mode, technologies like virtual learning help to facilitate convenient learning environments without the need to leave home and there is flexibility to fit learning into hectic schedules.

Other than pan-ASEAN initiatives like the transport agreements, there are also specific sub-sectors of tourism that need attention, for example, the need to reduce piracy, cross-boundary crimes and maritime armed robberies so that the regional ASEAN cruise industry can take off.[4] In the field of education, particularly for Less Developed Countries (LDCs) and economies, technical and vocational education is just as important as university education. There is a need to train more technically skilled staff who can operate the nuts and bolts of an industrializing economy just as much as the need to cultivate managers with more theoretical and management skills.

---

[2] *Ibid.*, p. 52.

[3] NEAT Working Group, NEAT Working Group on Enhancing People to People Connectivity -Education, Tourism and Cultural Exchange-Final Report Tokyo 29 July 2014 (Japan: NEAT), 2014, p. 2.

[4] ASEAN Secretariat, *op. cit.*, p. 55.

The Regional Tourism Industry: Besides local implications, heritage tourism and eco-tourism also take on a regional portfolio in the ASEAN context. Given changing trends within the tourism industry and also to tap into intra-ASEAN potential for internal tourism, the terminology adopted by the organization to encompass heritage, cultural and environmental tours using local community resources and assets is "community-based tourism."

The ASEAN website adopts and uses an NGO definition of community-based tourism: "Tourism Concern of the United Kingdom describes community-based tourism as that which aims to include and benefit local communities, particularly indigenous peoples and villagers. Community tourism projects should give local people a fair share of the benefits/profits of tourism and a say in deciding how incoming tourism is managed."[5]

Three points are important here. First, the definition of community-based tourism indicates a people-centred approach to running local tours. The definition exceptionalizes native-born individuals and dwellers of rural regions. In other words, the focus is on the most economically vulnerable but culturally richest individuals still engaged with traditional practices. The second feature is the idea of equitable distribution of tourist revenue from the use of community cultural or natural resources. The third point is the allocation of power and management control to the local denizens residing at that resource.

The advantages of involving local residents are many. They include the fact that local communities tend to understand their resources most optimally, acquainted with the local climatic, geological or geographical conditions. They are also most familiar with the local customs of the land, steeped in the traditional practices of their ancestors. Inevitably, when it comes to cultural assets and preservation of those resources, one has to include issues of identity formation, sustainability and cultural attachment to those resources.

These three issues give rise to the importance of conservation. The items to conserve, selective narratives and interpretations of local histories behind those material artefacts or memory-based constructs are dependent on the conservation authorities and co-option, resistance or collaboration from the local community. Working with local communities

---

[5]ASEAN, "Community-Based Tourism" in the ASEAN Tourism website (downloaded on 7 June 2016). Available at http://www.aseantourism.travel/explore/sub/culture-and-heritage/community-based-tourism.

takes on a new meaning when national sovereignties are juxtaposed and superimposed onto a regional landscape. On 31 December 2016, the ASEAN Economic Community (AEC) quietly came into being.

At present, within the ASEAN Community, in terms of Community-Based Tourism (CBT), each ASEAN country has come up with a list of three heritage spaces that they wish to promote under CBT. There are many benefits of an enhanced tourism industry in ASEAN. It can be an important source of foreign exchange for individual countries, particularly those least developed countries (LDCs) or economies. The heritage and tourism sites can be useful for creating awareness of gender and racial minority issues through material artefacts and physical landscapes/structures. Heritage tourism, particularly natural heritage resources, can also create awareness of the importance of sustainability because overuse or overexploitation of natural resources may lead to issues like soil erosion or coastal area degradation which in turn can attract natural disasters that are harmful to local communities in the long run. China can help in this case by relaying its experience in the area of environmental clean-up to ASEAN countries.

China's Contributions to ASEAN People-to-People Connectivity: China's expertise offered through the BRI and AIIB will be useful for the tangible aspects of ASEAN aspirations for people-to-people connectivity. Funding, technology and expertise like building HSRs can help physically connect regions and disparate areas like far-flung islands. Tourism is another bridge between China and ASEAN. The Thai ambassador to China noted, in 2014, 11.4 million Chinese travelled to Southeast Asia, and 6.2 million people from ASEAN visited China, promoting people-to-people connectivity.[6]

An Inclusive, Loose and Open Process for Cooperating with Other Major Countries: Ultimately, ASEAN people-to-people connectivity may benefit from Chinese capacity-building knowledge and assistance, but the process is inclusive and so this learning and collaboration process can be widened to include other entrants. For example, in the tertiary

---

[6]Niyom, Theerakun, "ASEAN builds a community, enhances regional cooperation" dated 9 August 2015 in the Global Times (downloaded on 1 June 2016). Available at http://www.globaltimes.cn/content/936194.shtml.

education sector, the US is still leading in this sector. Therefore, indeed both China and ASEAN included can benefit from studying the best practices of leading US tertiary institutions and utilizing US standards for research in university reforms and upgrading.

This chapter ends the section on China and leads to this volume's following section on Hong Kong, a leading tech hub and SAR of China.

Bibliography

ASEAN, "Community-Based Tourism" in the ASEAN Tourism website (downloaded on 7 June 2016). Available at http://www.aseantourism.travel/explore/sub/culture-and-heritage/community-based-tourism.

ASEAN Secretariat, Masterplan on ASEAN Connectivity (Jakarta: ASEAN Secretariat), 2011, p. 18.

NEAT Working Group, NEAT Working Group on Enhancing People to People Connectivity-Education, Tourism and Cultural Exchange-Final Report Tokyo 29 July 2014 (Japan: NEAT), 2014, p. 2.

Niyom, Theerakun, "ASEAN builds a community, enhances regional cooperation" dated 9 August 2015 in the Global Times (downloaded on 1 June 2016). Available at http://www.globaltimes.cn/content/936194.shtml.

Part IV

Hong Kong and its Tech Sector

Chapter 7

The Tech Hub of the Hong Kong Special Administrative Region (SAR)

Within China, the pioneer region in the capitalist experiment, is the tiger economy of Hong Kong SAR which had been attracting tech companies to its shore by building robust R&D facilities, providing infrastructure support for the tech industry and maintaining a track record for IPR protection.[1] The Hong Kong state agencies like InvestHK's start-up survey noted an increase of 51% in 2020 in the number of start-ups (especially Fintech, e-commerce/supply chain management/logistics technology, information, computer and tech start-ups among the data-related industries).[2] From 2017 before the pandemic and Hong Kong authorities achieved this by organizing domestic organisations to construct robust local and global partnerships.[3] Hong Kong's Fintech activities include data analytics, robotics, big data, P2P, natural language processing, foreign exchange services, cybersecurity, e-commerce, wealth management, robo-advisory in financial services and focused platforms to improve communications/collaborations for Fintech participants/public (e.g. The Fintech Facilitation Office of

---

[1] Hawksford, "Why Is Hong Kong a Unique Place for Tech Companies?" dated 7 April 2021 in GuideMeHongKong Hawksford (downloaded on 7 April 2021). Available at https://www.guidemehongkong.com/in-the-news/2021---why-is-hong-kong-a-unique-place-for-tech-companies.

[2] *Ibid.*

[3] *Ibid.*

Hong Kong Monetary Authority HKMA), Fintech Contact Point (Securities and Futures Commission SFC) and Insurtech Facilitation Team (Insurance Authority IA)).[4]

In 2022, Hong Kong signalled that it essentially wants to encourage innovative companies seeking IPOs on their bourse to enter at a lower threshold in revenue. The Hong Kong bourse is arrowing cutting-edge technology and eco-green firms (including those with zero income) in their drive to increase its liquidity pool for accelerating tech advancement after public consultation for two months before implementation.[5] The HKSE will delete revenue pre-conditions for pre-commercial companies[6] with market valuation pegged minimally at HK$15 billion (US$1.91 billion) while it will mandate commercial companies with valuations at least at HK$8 billion with profits at no less than HK$250 million.[7] CEO of HKEX Nicolas Aguzin articulated: "We expect the new proposed specialist technology rules will help to drive growth in talent and investment across these five frontier industries, such as in green technologies, in the region and beyond. These new proposals will expand the range of companies that can access Hong Kong's deep, liquid, international markets and will offer investors even greater choices."[8]

Under the new regime, retail investors can trade shares of these companies while institutional investors will be allocated minimally 50% of the overall shares offered in the IPO process, as Bonnie Chan, head of listing at HKEX, explained:

---

[4]Fung, Doris, "Innovation and Technology Industry in Hong Kong" dated 6 September 2022 in HKTDC Research (downloaded on 6 September 2022). Available at https://research.hktdc.com/en/article/MzEzOTIwMDIy.

[5]Zhang, Tianyuan, "HK to facilitate high-tech floats" dated 21 October 2022 in China Daily (downloaded on 21 October 2022). Available at https://www.chinadaily.com.cn/a/202210/21/WS6351f6c0a310fd2b29e7dbe9.html.

[6]Pre-commercial companies is defined as companies mainly involved in research and development (R&D) with income not at the level of HK$250 million. *Source*: Zhang, Tianyuan, "HK to facilitate high-tech floats" dated 21 October 2022 in China Daily (downloaded on 21 October 2022). Available at https://www.chinadaily.com.cn/a/202210/21/WS6351f6c0a310fd2b29e7dbe9.html.

[7]Zhang, Tianyuan, *op. cit.*

[8]*Ibid.*

We took into account the unique features of specialist technology companies, in particular, the uniqueness of the role technology plays in their business and their early stage of development relative to other listing applicants. We believe our proposed rules strike the right balance between upholding market quality and creating a commercially viable chapter that meets the fundraising needs of the leading companies of tomorrow.[9]

Infrastructure Facilities in Hong Kong

Government-owned Cyberport is a creative digital platform with 1,500 start-ups and tech firms with stated objectives to be a ranking international I&T hub, improving the domestic economy by cultivating digital start-ups and entrepreneurs, motivating joint business opportunities and fast-tracking digitalization through strategic schemes and collaborations.[10] The Hong Kong Cyberport Management Company Limited (Cyberport) is a 47,000 square feet facility that hosts Fintech firms and MNCs like Microsoft, Lenovo and IBM and provides financial investment services, mentorship, business advisory, publicity/advertising, marketing/promotion, alumni networks and business development for its tenants.[11]

Hong Kong's start-ups expanded by 12% to 3,755 hiring more than 13,000 people in 2021.[12] But it is not resting on its laurels. The Cyberport Creative Micro Fund (CCMF)'s HK$100,000 cash grant and access to Cyberport resources for innovators and entrepreneurs provide HK$100,000 grants for tech start-ups to finance activities related to product development within six months. The grant is provided in three stages: (Stage 1) HK$10,000 dispensed after inking the CCMF agreement, (Stage 2) HK$45,000 after submission of the interim report and (Stage 3) providing HK$45,000 upon completion and final report submission.[13] The conditionalities for this grant must meet the following criteria: Hong Kong ID above 18 years old, firms registered and limited incorporated in Hong Kong.[14]

---

[9] *Ibid.*
[10] Fung, Doris, *op. cit.*
[11] Hawksford, *op. cit.*
[12] Fung, Doris, *op. cit.*
[13] Hawksford, *op. cit.*
[14] *Ibid.*

Through the 2017 HK$2 billion Innovation and Technology Venture Fund (ITVF) for cross-investments in domestic innovation and technology start-ups matched by private venture capital funds, the Hong Kong authorities are attracting "smart money" to augment investments in domestic I&T start-ups (HK$170 million in 23 start-ups and raised HK$1.5 billion of private investment by end of 2021).[15] Hong Kong start-ups zero in on the following technologies: information and communication technologies (ICT), software as a service (SaaS), the IoT, data analytics, biotech, AI, robotics, virtual reality (VR), AR and new materials with applications forwarded in the areas of Fintech (financial technology), smart city and smart home, healthcare and big data applications.[16]

These niche research areas certainly need their own specialized facilities. Hong Kong ranked 6th in tech infrastructure out of 132 economies in the 2021 Global Innovation Index and, from the 1998 I&T development blueprint onwards, the authorities invested an initial HK$5 billion Innovation and Technology Fund (ITF) to set up Hong Kong Applied Science and Technology Research Institute (ASTRI), Science Park, Cyberport and five R&D centres.[17] By the end of 2021, ITF financed 27,466 projects with HK$26.5 billion in foundational industries (24.7% of these approved projects), IT (21.8%), electrical/electronics (21.3%) and biotechnology (15.2%).[18]

The 22-hectare Hong Kong Science and Technology Parks Corporation (HKSTP) hosts cutting-edge labs and facilities that lessen capital investment of tech firms in product design and development to low entry costs of new products into the consumer market and provide three incubation programs, such as Incu-App, Incu-Tech and Incu-Bio.[19] The Hong Kong Science Park hosts 1,100 tech firms and more than 11,000 R&D stakeholders to build a leading I&T ecosystem in Hong Kong with the authorities providing cutting-edge equipment, R&D software and test-bed platform while encouraging work interactions, investors' networking, business development and commercialisation of products.[20] Hong Kong

---

[15] Fung, Doris, *op. cit.*

[16] *Ibid.*

[17] *Ibid.*

[18] *Ibid.*

[19] Hawksford, *op. cit.*

[20] Fung, Doris, *op. cit.*

authorities have started Phase 2 Science Park Expansion Programme and Cyberport 5 (28,000 and 63,000 square metres respectively mainly for R&D/I&T firms) while Tseung Kwan O Industrial Estate (TKOIE) Data Technology Hub (DT Hub, facilitation of data transfer operations and global telecommunications at data and switching centres) has started operating in the 4th quarter of 2020.[21]

Next to Science Park, InnoCell was constructed at the end of 2020 with 500 residential spaces (including flexible design and shared workspaces for Science Park research staff) while the TKOIE Advanced Manufacturing Centre in TKOIE and Yuen Long Microelectronics Centre are in the immediate pipeline (with 140,000 square metres for smart production and high-end manufacturing).[22]

Biomedical Industry

A two-year Incu-App dispenses up to HK$860,000 worth of R&D support, expert mentorship, industrial content development, application platforms, market support and value chains through partnerships with leading firms in Hong Kong.[23] Besides the private sector, Hong Kong is also working with leading tertiary institutions and research organs overseas. They include Cornell University-CityU 2010 pioneering academic programme for veterinary medicine, MIT overseas Innovation Node in Hong Kong 2016 with "ready access to a unique manufacturing infrastructure that encourages prototyping and scaleup" and Sweden's Karolinska Institutet Science Park research facility for regenerative medicine/stem cell (Swedish pioneering overseas centre in 2016).[24] Incu-Tech offers up to HK$1.29 million for financial help for 3 years to assist start-ups focused on deep tech innovation research, working with mentors, strategic

---

[21] Hong Kong authorities has started Phase 2 Science Park Expansion Programme and Cyberport 5 (28,000 and 63,000 square metres respectively mainly for R&D /I&T firms) while Tseung Kwan O Industrial Estate (TKOIE) Data Technology Hub (DT Hub, facilitation of data transfer operations and global telecommunications at data and switching centres) has started operating in 4th quarter of 2020.

[22] Fung, Doris, *op. cit.*

[23] Hawksford, *op. cit.*

[24] Fung, Doris, *op. cit.*

partners, investors and experts recommended by the appointed Incu-Tech account manager.[25]

Mainland institutions are also working strategically with Hong Kong research and tertiary institutions, funded by public funds and investments to further life science and biotechnology work: Hong Kong Institute of Biotechnology (HKIB, established in 1988) and Biotechnology Research Institute (BRI). Founded in 1988, it carries out R&D, tech transfers, product commercialization, a Biotechnology Incubation Centre for domestic and global start-ups/firms and other forms of equipment, technical and adminis-trative support.[26] As a large-scale clinical trial centre for new drugs (includ-ing those from the US) through its two Phase I Clinical Trial Centres established in 2014, Hong Kong has strong cooperation with multinational pharmaceutical companies and National Medical Products Administration,[27] with HK Eye Hospital, HKU and CUHK as the first three hospitals outside mainland China to trial drugs for registration in China.[28]

In 2017, Guangzhou Institutes of Biomedicine and Health established a stem cell and regenerative medicine research facility at the Science Park.[29] Collaboration with Shenzhen is also strengthening. Hong Kong has two InnoHK Research Clusters at Science Park (Health@InnoHK on healthcare tech and AIR@InnoHK on AI/robotics with 28 research labs) and, due to the focus on life/health R&D after COVID-19 pandemic, the 2022–2023 Budget allocated HK$10 billion to reinforce industrial chains and establish InnoLife Healthtech Hub in Hong Kong–Shenzhen Innovation and Technology Park.[30]

The four-year Incu-Bio provides a maximum of HK$6 million in financial aid operated by the Biomedical Technology Cluster to provide help for biomedical tech start-ups in Hong Kong through networking with investors and policy stakeholders.[31] One of the most important stakehold-ers is the Institute of Higher Learning and Research in Hong Kong. Hong Kong universities publish 250 high-impact biomedical publications,

---

[25] Hawksford, *op. cit.*

[26] Fung, Doris, *op. cit.*

[27] Formerly China State Food and Drug Administration.

[28] Fung, Doris, *op. cit.*

[29] *Ibid.*

[30] *Ibid.*

[31] Hawksford, *op. cit.*

participate in large international genomic projects and discovery of emerging infectious diseases (e.g. SARS and avian flu virus), license patents from its start-ups and host ISI Essential Science Indicators-rated medical institutions (with results like molecular diagnostics utilizing cell-free DNA in blood, "Oral Arsenic Trioxide" for treat Acute Promyelocytic Leukemia (APL)).[32]

In the 2018–2019 Budget, Hong Kong authorities focused on biotech, AI, smart city and Fintech while providing HK$130 billion for 8 I&T areas (R&D, tech talents, investment funding, tech research infrastructure, legislations/regulations reviews, opening up state data, augmented procurement arrangements and popular science education).[33] Smart city development is guided by the authorities' blueprint. Hong Kong's Smart City Blueprint emphasizes 6 I&T sectors: (1) smart mobility (intelligent transport system/traffic management), (2) smart living (free public Wi-Fi and eID for public/private sector services), (3) smart environment (green buildings, energy efficiency and waste management/pollution tracking), (4) people capacity-building, (5) smart government (open data, smart city infrastructure and e-public service) and (6) smart economy (sharing economy).[34]

Supported by Innovation and Technology Commission and the Hong Kong government, there are also 16 financial programmes under the ITF for R&D, technological implementation, tech talent development, reinforcing tech cultures and tech start-ups.[35] The "Technology Start-up Support Scheme for Universities" (TSSSU) established by Innovation and Technology Commission (ITC) in 2014 dispenses yearly funding (maximum HK$8 million but doubled in the 2022–2023 Budget to HK$16) to six local universities for spurring tech start-ups (with each getting a maximum of HK$1.5 million annually for a limit of 3 years).[36]

To strengthen Hong Kong companies' tech drive, the Technology Voucher Programme (TVP) subsidies were instituted to help domestic companies utilize technological services and solutions to enhance

---

[32]Fung, Doris, *op. cit.*

[33]*Ibid.*

[34]*Ibid.*

[35]Hawksford, *op. cit.*

[36]Fung, Doris, *op. cit.*

productivity or improve their business operations.[37] The Hong Kong authorities have set up specialized institutions for enhancing productivity. The Hong Kong Productivity Council (HKPC) is a multidisciplinary institution to augment productivity and global competitiveness of Hong Kong companies through the market-friendly applied R&D in smart products, smart manufacturing, automation, new materials, surface treatment, smart mobility, green transportation, environmental tech, consultancy services, tech transfers, training and other services.[38] As for the TVP, each applicant may get cumulative financing up to a maximum of HK$600,000 and up to six projects, so each individual project is to be completed at a time within 12 months and awardees must make up at least 25% of the overall project expenditure.[39]

By the end of January 2022, 48 healthcare and biotechnology firms were listed in Hong Kong, raising more than HK$110 billion overall, becoming Asia's biggest and globally second-largest biotech fundraising location and the Central Government allowed pre-profit HK-listed biotech firms and mainland-listed stocks on the Sci-Tech Innovation Board meeting pre-requisites to have mutual market access between the two financial markets.[40]

Concluding Remarks

Some of the future challenges may include whether Chinese hybrid centralized socialist political systems co-existing with a market-oriented economy but in the process of recentralizing state supervision can remain compatible with the Western capitalist economic system. The following chapter examines Hong Kong's GBA macro-region to contextualise its techno-economic future. The nexus between Hong Kong and China's development is perhaps visibly clearest here.

---

[37] Hawksford, *op. cit.*
[38] Fung, Doris, *op. cit.*
[39] Hawksford, *op. cit.*
[40] Fung, Doris, *op. cit.*

Bibliography

Fung, Doris, "Innovation and Technology Industry in Hong Kong" dated 6 September 2022 in HKTDC Research (downloaded on 6 September 2022). Available at https://research.hktdc.com/en/article/MzEzOTIwMDIy.

Hawksford, "Why Is Hong Kong a Unique Place for Tech Companies?" dated 7 April 2021 in GuideMeHongKong Hawksford (downloaded on 7 April 2021). Available at https://www.guidemehongkong.com/in-the-news/2021---why-is-hong-kong-a-unique-place-for-tech-companies.Zhang, Tianyuan, "HK to facilitate high-tech floats" dated 21 October 2022 in China Daily (downloaded on 21 October 2022). Available at https://www.chinadaily.com.cn/a/202210/21/WS6351f6c0a310fd2b29e7dbe9.html.

Chapter 8

Greater Bay Area Technological Development

Introduction

The GBA is designated as a national champion for tech innovation by the central government of the PRC by augmenting the strengths of its cities in the areas of technology, finance, shipping, high-tech manufacturing, hospitality, supply chains and talent development.[1] Articulated in the state-owned media *China Daily*, China hopes the Guangdong-Hong Kong-Macao GBA can transform into a new engine powering China's economic growth with more efforts by all stakeholders in industrial upgrades, technological innovation and augmenting its latecomer advantage.[2] Guan Qingyou, Dean and Chief Economist at Reality Institute of Advanced Finance indicated the Bay Area is a potential market for China's dual circulation

---

[1] Ernst and Young (EY) Greater China, "Strategy for a tech-driven Greater Bay Area" dated 11 June 2021 in EY (Ernst and Young website) (downloaded on 11 June 2021. Available at https://www.ey.com/en_cn/china-opportunities/strategy-for-a-tech-driven-greater-bay-area.

[2] Zhou, Mo, "GBA seen as new engine driving China's economic growth" dated 22 September 2020 in China Daily (downloaded on 22 September 2020). Available at https://www.chinadailyhk.com/article/144097#GBA-seen-as-new-engine-driving-China's-economic-growth.

151

policy where internal economic circulation complements the other track that focuses on external markets for China's future development.[3]

GBA is creating an eco-system for tech and innovation through more public/private investments, regulatory development, talent mobility and preferential tax treatments/incentives for R&D firms to attract corporate investments.[4] The central government's support is essential for the GBA's technological development to take off. For example, China's 14th FYP (2021–2025) backs up Hong Kong's plans to evolve into a global innovation and technology hub.[5] Strong enterprise sponsorships also spawned from favourable government policies arising from the National Intellectual Property Strategy in 2008, its ambitious national/regional targets/quotas for patent applications/ownership, e.g. regional online firms are proactively using their corporate venture capital to make investments on promising venture, driven by state incentives for patent applications.[6] In October 2020, China's Standing Committee of the NPC promulgated legal amendments to the Patent Law for implementation in June 2021 in the areas of partial design protection, design protection period increase to 15 years, higher statutory damages and punitive damages, potentially benefiting the pharmaceutical industry.[7]

The state also instituted venture capital (VC) funding promotion policies like co-financing private VC funds from local government funds and, with a robust presence in Shenzhen, China's VC market grew globally to the second biggest market.[8] There appears to be general optimism about GBA tech and innovation growth among the companies based in that sub-region. A KPMG study indicated the following results. 61% of companies in GBA cities have expansion plans, 79% of tech companies opined tech/innovation sector benefits from the GBA and 67% of

---

[3] *Ibid.*

[4] KPMG, "Greater Bay Area Update" in KPMG Tech and Innovation GBA newsletter dated February 2021 (downloaded on 1 March 2021). Available at https://assets.kpmg/content/dam/kpmg/cn/pdf/en/2021/02/greater-bay-area-update-tech-and-innovation-gba-newsletter.pdf, p. 6.

[5] *Xinhua*, "GBA offers huge potential for health tech investment" dated 24 November 2021 in China Daily (downloaded on 24 November 2021). Available at https://www.chinadailyhk.com/article/248942.

[6] Ernst and Young (EY) Greater China, *op. cit.*

[7] KPMG, *op. cit.*, p. 3.

[8] Ernst and Young (EY) Greater China, *op. cit.*

tech/innovation firms surveyed projected increased R&D spending in 2021–2024 in another GBA city.[9] It also indicated 92% of tech executives have a talent recruitment/investment strategy in 2021–2022 and 53% of tech/innovation executives project business revenues increase of at least 30% in 2021–2024.

Guangdong

In terms of tech infrastructure, Guangdong province has been proactive in funding cutting-edge infrastructure useful for the tech industry. Guangdong province is putting funding into 5G networks with 41,000 5G base stations, given that the value of Guangzhou's tech contract deals came up to approximately US$14.9 billion in 2019 and the province has already sunk US$2.2 billion on the development of deep-sea ecosystem, a dynamic wide-range velocity vehicle test unit, etc.[10]

The Guangdong Bureau of Statistics was commissioned by the central government to research the features of nine GBA cities' technological sector and develop a GBA innovation development index focused on six sectors: R&D, patent development and product output, R&D manpower, enterprise capital expenditure, technological upgrade investment and the innovative environment.[11] The market environment for innovation development in the nine GBA mainland cities multiplied by three times in 2013–2018 generating tech innovation breakthroughs, facilitated by accelerated increase in R&D firms, patents/new product output (that tripled in that period).[12] This resulted in Guangdong's GBA cities getting ahead of the rest of the nation in global Patent Cooperation Treaty (PCT) applications with 24,700 patent applications covering 40% of the country's total in 2019 and the GBA also hosts approximately 20.5% of Chinese companies with R&D facilities.[13]

The industry incrementally created an ecosystem by forming networks of new and veteran talents in the appropriate sectors and then

---

[9]KPMG, *op. cit.*, p. 2.
[10]KPMG, *op. cit.*, p. 5.
[11]Ernst and Young (EY) Greater China, *op. cit.*
[12]*Ibid.*
[13]*Ibid.*

motivating them to have fruitful interactions, and Shenzhen is considered the strongest GBA city in enterprise capital expenditure on R&D, equipped with a large-scale technology sector and its private firms, R&D human resource and an innovative environment.[14] Zhuhai has well-developed research-oriented manpower, an environment in innovation and a niched biomedical sector that differentiates it from Shenzhen's industrial niched in electronics and communication equipment and Dongguan, a comparatively later entry into the innovation and technology developmental sector and a robust patent/new product output that is catching up with Shenzhen.[15]

Located between Shenzhen and Guangzhou, Dongguan is a smart manufacturing/R&D centre. It has a 4th National Comprehensive Science Centre, two science parks (Dongguan Songshan Lake Science City and Shenzhen Guangming Science City), Huawei innovation/production platforms, 5,700 high-tech firms, Neutron Science Center of Dongguan (operational since 2018), China's sole spallation neutron source (4th globally), a new materials research/industrial cluster for new materials, microparticles and biomedicine.[16] Zhuhai development started in 2012 with state plans to move its industrial sector up the value chain from processing industries using tax concessions, R&D subsidies and seminars/exhibitions to assist the private sector and SMEs in technology.[17]

The current COVID-19 coronavirus pandemic has created conditions for catering to new forms of manufacturing processes. The COVID-19 coronavirus pandemic has created the demand for smart manufacturing and e-commerce that harness data analytics, robotics and IoT to manufacture personalized/customized complex goods for Original Equipment Manufacturer (OEMs) for export markets like emerging ASEAN.[18] In May 2020, Chinese regulators came up with 26 measures to promote the financial services in the GBA with tech innovation as a major focus; the document included deepening Fintech cooperation, constructing a blockchain trade financing information service platform, supporting

---

[14] *Ibid.*
[15] *Ibid.*
[16] KPMG, *op. cit.*, p. 5.
[17] Ernst and Young (EY) Greater China, *op. cit.*
[18] KPMG, *op. cit.*, p. 4.

research on innovative technologies and expanding cross-border electronic payments.[19]

In terms of challenges and weaknesses, the same study indicated laggard software in place in the GBA. The GBA's softer infrastructure like R&D, manpower attraction/training, enterprise capital funding for R&D as a component of value-added of industry (VAI) and local government finances and general research environment for innovation/tech development are trying to keep up with the infrastructure and financing progression.[20] The study identified the reason for laggard progress in software development. It is mainly due to the fact that economic growth in the tech areas has surged ahead rather than the industry player/authorities failing to put efforts into developing them as the GBA is already ahead with 4 of its 10 leading cities in China leading in this industry due to exponential increase in R&D spending.[21]

Changes in HK are fast and furious. Institutional investors in HK/Macau are permitted to engage in GBA private equity investment/venture capital funds. They can do so through Qualified Foreign Limited Partners (QFLPs), involving the state, entrepreneurs, financial industry and technology talents, with these stakeholders urging the state to subsidize patent applications, provide R&D facilities, add R&D fiscal funding, encourage flow of ideas and institute tax cuts/regulation reforms.[22]

The Special Position of Hong Kong

The GBA is keen on Innovation and technology (I&T) development to improve quality of life and create quality jobs for young individuals and the Hong Kong authorities have identified their comparative advantages in technological R&D, its status as a global finance centre and the ability to aggregate GBA resources.[23] For example, Alfred Sit Wing-hang,

---

[19] *Ibid.*, p. 4.

[20] Ernst and Young (EY) Greater China, *op. cit.*

[21] *Ibid.*

[22] *Ibid.*

[23] Constitutional and Mainland Affairs Bureau, "Innovation and Technology" dated 2018 in the Greater Bay Area website (downloaded on 1 January 2022). Available at https://www.bayarea.gov.hk/en/opportunities/it.html.

Secretary for Innovation and Technology of the HKSAR, articulated at the Asia Summit on Global Health in Hong Kong in November 2021 that HK has a good future in healthcare tech development. He said HK has many global talents in the life/health sector, a strong research backdrop and a lively and complementary ecosystem that encourages health tech-related start-ups to set up shop: "Currently, we are Asia's largest and the world's second-largest fundraizing hub for biotech companies."[24]

In the Chief Executive 2020 Policy Address in November 2020, then CE Carrie Lam proposed a five-year, HKD 2-billion "Global STEM Professorship Scheme" to lure top R&D foreign talent to Hong Kong, create 2000 jobs in the I&T sector for HK graduates to work in the GBA through the GBA Youth Employment Scheme.[25] This can create more enthusiasm in STEM subjects among undergraduates to join R&D and commercialization initiatives and, in October 2020, Hong Kong Science and Technology Park collaborated with the City University of Hong Kong to construct a global-standards Good Manufacturing Practice (GMP) facility to work on cancer therapies, regenerative medicine and related biomedical technologies.[26] Cross-border fast-track customs-clearance features can accelerate GBA exchanges of scientific research samples, lab reagents and genetic resources to facilitate the innovation environment and augment growth in biomedicine, high-end medical equipment and DNA genetic testing and further strengthen Health@InnoHK healthcare tech research cluster in the Science Park.[27]

Hong Kong and Macau have already developed niches of their own as a financial hub and recreational hub, respectively. They have the potential to complement the tech industry. In 2016, the Hong Kong Monetary Authority (HKMA) established the FinTech Supervisory Sandbox (FSS) for banks and their tech company partners to carry out pilot trials of their Fintech initiatives, implement creative financial products in a controlled environment with actual clients to acquire real-life experience, motivate regulatory authorities/government to evaluate the implementation of the regulations.[28]

---

[24] Xinhua, *op. cit.*

[25] KPMG, *op. cit.*, p. 3.

[26] *Ibid.*, p. 3.

[27] *Ibid.*, p. 7.

[28] *Ibid.*, p. 9.

Hong Kong's policy efforts also extend to the scientific community. According to *Global Scientific and Technological Innovation Centers Evaluation Report* published by the Shanghai Information Center, Hong Kong is 15th compared to Shenzhen's 27th and Guangzhou's 46th in 2020 in innovativeness, with HK enjoying strong fundamental research (benchmarked by academic publication citations, global ranking of its universities, scientific research and research facilities), potentially producing human talents for the GBA.[29]

The Hong Kong government is increasing efforts in I&T development in eight major areas with a US$100 billion investment, including increasing resources for R&D, pooling together technology talent, providing investment funding/technological research infrastructure, reviewing legislations and regulations, opening up government data, enhancing procurement arrangements and augmenting popular science education.[30] Hong Kong is also counting on their universities being in the top 100 Quacquarelli Symonds (QS) World University Rankings and they have five of them in the 2020 ranking while the Science Park is pushing forward with two InnoHK research clusters on healthcare technologies, artificial intelligence (AI) and robotics.[31]

Through these institutions, Hong Kong wants to bring in world-class research institutions and tech firms to carry out R&D projects in partnership with their universities and research institutions to form a cluster that will gather/develop local talents with a third InnoHK research cluster specifically targeted at international R&D efforts.[32] According to Donald Li, a member/national political advisor of the 13th National Committee of the Chinese People's Political Consultative Conference (CPPCC) and President of the World Organization of Family Doctors, in the healthcare industry, for example, HK has globally renowned universities with outstanding health/medicine and medical departments that have turned out graduates with globally recognized achievements.[33] In July 2020, HKSTP's InnoAcademy launched the Technology Leaders of Tomorrow Programme for science/engineering graduates to work for two years in a full-time R&D occupation at one of the 11 HKSTP partner firms,

---

[29] Ernst and Young (EY) Greater China, *op. cit.*
[30] Constitutional and Mainland Affairs Bureau, *op. cit.*
[31] *Ibid.*
[32] *Ibid.*
[33] Xinhua, *op. cit.*

experiencing innovation leadership training, career guidance, networking and taking part in an innovation & technology (I&T) industry
boot-camp.[34]

The Hong Kong government and the Chinese Academy of Sciences
(CAS) inked a memorandum of understanding (MOU) in November 2018
where CAS sets up an affiliated institution for its Guangzhou Institutes of
Biomedicine and Health and Institute of Automation's R&D programmes.[35] According to Donald Li, a member/national political advisor
of the 13th National Committee of the Chinese People's Political
Consultative Conference (CPPCC) and President of the World Organization
of Family Doctors, in the healthcare industry, for example, HK can use
some of the newest cutting-edge pharmaceutical drugs and prescribe/offer
such medicines.[36] Hong Kong has over 40 academicians from the CAS
and Chinese Academy of Engineering as part of a GBA academician alliance in Hong Kong with the affiliated institution of CAS in Hong Kong
becoming the secretariat for the alliance and supporting its daily
activities.[37]

China's 14th FYP (2021–2025) backs Hong Kong's innovation and
technology hub and it envisions Hong Kong and its comparative strengths
in many areas as a gateway to access GBA economic potential.[38] In
September 2018, the national Ministry of Science and Technology and the
Innovation and Technology Bureau of HKSAR inked an agreement to
promote I&T exchanges and they also started the Mainland-Hong Kong
Joint Funding Scheme in April 2019 to motivate scientific research cooperation between Hong Kong and different Chinese provinces.[39] Hong
Kong's tertiary/research institutions enjoy science/tech funding from the
central/provincial/municipal governments (plus relevant HK funding)
and, from mid-2019, HK Ministry of Science and Technology and state
agencies of Guangdong/Shenzhen can remit approved funds across the
border while HK academic institutions/hospitals or their mainland

---

[34] KPMG, *op. cit.*, p. 4.

[35] Constitutional and Mainland Affairs Bureau, *op. cit.*

[36] *Xinhua, op. cit.*

[37] Constitutional and Mainland Affairs Bureau, *op. cit.*

[38] *Xinhua, op. cit.*

[39] Constitutional and Mainland Affairs Bureau, *op. cit.*

offshoots can export human genetic resources to HK under certain conditions.[40]

Technology firms can reach into HK's financial market to invest in GBA R&D, e.g. in the *Opinion on Financial Support for the Construction of the Guangdong-Hong Kong-Macao Greater Bay Area* published in May 2020, private equity funds in HK/Macau were urged to fund cross-boundary financing in GBA innovative enterprises and to lists qualifying innovative firms in HK.[41] The HK government has put in place enhanced tax deduction for R&D expenditure with the initial US$2 million eligible R&D expenditure enjoying a 300% tax deduction and the remainder at 200%.[42] Externally, Hong Kong also has much to offer for other urban city developments. When it comes to constructing smart cities, ASEAN cities study Hong Kong ranked among the globe's top smartest cities, in terms of trendsetting, quality and creativity, especially when it comes to the utilization of geographic information systems (GIS) in urban planning.[43] Thus, Hong Kong and other GBA-leading cities can focus on these comparative advantages in smart city development and best practices for consultancy work with other urban city developments outside the GBA.[44]

Hong Kong and Shenzhen Team-Up

With more dependence on domestic consumption to drive Chinese economic growth, the Chinese leadership want Shenzhen's reform to continue unabated. On the 40th anniversary of the founding of Shenzhen Special Economic Zone (SEZ), President Xi Jinping ordered the National Development Reform Commission (NDRC) to ensure Shenzhen's autonomy through instituting 27 reforms and 40 authorized initiatives declared on 18 October 2020 in six sectors: market allocation of factors of production, business activities, technology and innovation and further opening-up.[45] These initiatives back Shenzhen to muster legislative power in the sectors of artificial intelligence (AI), big data, electric vehicles, drones

---

[40] *Ibid.*

[41] Ernst and Young (EY) Greater China, *op. cit.*

[42] Constitutional and Mainland Affairs Bureau, *op. cit.*

[43] KPMG, *op. cit.*, pp. 3–4.

[44] *Ibid.*, pp. 3–4.

[45] *Ibid.*, pp. 1–2

and biotech, enhancing the flow of skilled manpower resources and the institution of a more open business environment.[46]

As part of the GBA initiative, Shenzhen is still being developed as a city for sustained experimentation with the reform and opening up of Chinese economy centred on the theme of innovation for the transformation of its innovation centres, universities, research institutes and innovative corporations.[47] With a positive business/entrepreneurial environment, cutting-edge tech firms like Tencent, Huawei, Ping An and DJI forged the "Shenzhen spirit" of innovation, inclusiveness and boldness, something Sheng Ruisheng (board secretary and brand director of Ping An Insurance Group) argued is embedded in his 32-year old company in the form of persistence in innovation and technological breakthrough innovations.[48] There are constant innovation tie-ups in Shenzhen with Chinese entities outside the city. In July 2020, Huizhou signed contracts worth US$714.0 million with Shenzhen Sinlikon Supply Chain Management and Hit Robot Group based in China to build an industrial estate centred on 5G telecom tech and an R&D centre.[49]

There had been plans put in place for raising funds for I&T start-ups in Shenzhen. In June 2020, ChiNext start-up board on the Shenzhen Stock Exchange made changes to regulations concerning rules of new share sales and instituted a registration-based initial public offering (IPO) system to align ChiNext closer to the standards of Shanghai Stock Exchange's STAR Market and bring about listings of quick-expanding and innovative tech firms and augment transparency in China's capital markets.[50] Established in 2009, ChiNext has more than 800 companies (market valuation of US$1.3 trillion as of August 2020) and it has been increasing transparency through an open/paperless application review process while relaxing trading restrictions and, in June 2020, the Chinese regulator declared official rules/regulations for the new IPO system and started accepting applications under the registration-based system.[51] With regulations loosening up for local/foreign talents in GBA cities like Shenzhen, companies are employing them from overseas to work remotely while

---

[46] *Ibid.*

[47] *Ibid.*

[48] Zhou, Mo, *op. cit.*

[49] KPMG, *op. cit.*, p. 5.

[50] *Ibid.*, p. 4.

[51] *Ibid.*

global companies are looking into the potential abundance of VC/PE capital for tech firms offered through financial institutions with mature tech funds specialized in loans to high-growth industries.[52]

President of the China Association of Small and Medium-Sized Enterprises (SMEs) Li Zhibin articulated that Shenzhen, with its sound business/entrepreneurial environments, has the Shenzhen government has regarded technological innovation as the "primary driver of economic development [and] primary growth point of economic development" since 1995.[53] One strategy is to pair off Shenzhen's leading innovative technologies as China's unofficial Silicon Valley with Hong Kong's global financial hub status, particularly in the areas of developing capital markets, enterprise R&D, expanding talent pool and tech ecosystem.[54] The Cross-boundary Wealth Management Connect Pilot Scheme in June 2020 can add to existing cross-GBA financial connections, especially in asset management, Fintech and insurance to augment HK's international financial and offshore RMB hub standing to enhance access to cross-border investments in GBA innovation and technology companies.[55]

HK/Shenzhen Municipal People's Government constructed HK-Shenzhen Innovation and Technology Park. It is the largest of its kind at the 87-hectare Lok Ma Chau Loop while mainland authorities co-developed Shenzhen-HK Innovation and Technology Co-operation Zone (made up of Lok Ma Chau Loop Hong Kong-Shenzhen Innovation and Technology Park and "Shenzhen Innovation and Technology Zone" in north Shenzhen River) to encourage the flow of talent, capital, goods, information.[56] Sixteen SKLs and six Hong Kong Branches of Chinese National Engineering Research Centres are conducting research on a spectrum of technological fields while foreign institutions such as Massachusetts Institute of Technology (MIT)'s first overseas innovation node and the Swedish Karolinska Institutet's first overseas research centre have set up operations in Hong Kong.[57]

---

[52]*Ibid.*, p. 6.
[53]Zhou, Mo, *op. cit.*
[54]Ernst and Young (EY) Greater China, *op. cit.*
[55]KPMG, *op. cit.*, p. 4.
[56]Constitutional and Mainland Affairs Bureau, *op. cit.*
[57]*Ibid.*

Hong Kong also put in place the fast-track Technology Talent Admission Scheme arrangement to get foreign/mainland talents and nurture local talents, while Smart City Blueprint for Hong Kong implemented in December 2017 intended to develop Hong Kong into a world-class smart city by implementing I&T for improving city management and quality of living and sustainable economic development.[58] The Blueprint included the six areas of "Smart Mobility," "Smart Living," "Smart Environment," "Smart People," "Smart Government," "Smart Economy" and 70 initiatives that are published in the Smart City Blueprint for Hong Kong 2.0 2020 report.[59] Targeted state policies will be assisting industry development. They do so by encouraging more private equity/venture capital in the tech sector, possibly increasing endowments of ranking Shenzhen tech research universities, coupled with November 2019 easing of restrictions for HK permanent residents (PRs) and facilitating expatriates to operate in mainland complemented by city government subsidies for attracting foreign talents through income tax, housing and transportation.[60] Interestingly, in terms of housing, Hong Kong is utilizing technologies to resolve the housing shortage and space constraints challenges.

Housing Technological Collaboration between Hong Kong and China: Made in Shenzhen Solutions[61]: Hong Kong has a hilly and mountainous terrain. Its tallest peak Tai Mo Shan stands at 957 metres and sits in the middle of the city.[62] A good 250,000 Hong Kong residents reside in subdivided apartments while the entire 7.4 million-strong population is squeezed into just 25% of the total land area of 426 square miles located along the contours of the Victoria Harbour; the majority of rural spaces are conserved green spaces that are under legal protection.[63] Seventy percent of Hong Kong's land is pristine rural and elevated terrains and 40%

---

[58] *Ibid.*

[59] *Ibid.*

[60] Ernst and Young (EY) Greater China, *op. cit.*

[61] The author first broached this topic in: Lim, Tai Wei, "Housing Policies in Hong Kong" dated 6 February 2020 in EAI Background Brief No. 1500 (Singapore: NUS EAI), 2020.

[62] Wong, Maggie, "Tai Mo Shan: How to tackle Hong Kong's highest peak" dated 7 February 2018 in CNN (downloaded on 7 February 2018). Available at https://edition.cnn.com/travel/article/tai-mo-shan-hong-kong/index.html.

[63] Su, Alice, "Hong Kong aims to solve its housing crisis with an $80-billion artificial island."

of its land area is conserved and protected (including marine parks).[64] Hong Kong government's 2016 study indicated that the territory requires a minimum of 2,965 acres of extra space for apartments, economic expansion and maintaining high lifestyle standards.[65]

In January 2019, US-based survey firm Demographia's international rankings listed Hong Kong as the world's least affordable housing market for nine continuous years.[66] The average price for a private housing in Hong Kong in 2019 was US$1,235,220.[67] The average price for a property in Hong Kong is 20.9 times the average yearly household income while in second-ranked Vancouver, a resident needs 12.6 years to accumulate income to purchase a property and in New York, it is 5.5 years.[68]

According to the Kwai Chung Subdivided Flat Residents Alliance survey, the average living space for Hong Kong's lowest income individuals was approximately 50 square feet, almost the same size as shared facilities in correctional facilities while nano flats (below 200 square feet) go for US$500,000.[69] Even car parks are premium spaces in Hong Kong. The average parking space in Hong Kong is approximately 135 square feet.[70] In 2017, a record was set when one parking lot was sold for a whopping US$664,000![71]

---

[64] Greening, Landscape and Tree Management Section Development Bureau, "Parks, Country and Marine Parks" dated 2016 in the Greening, Landscape and Tree Management website (downloaded on 1 January 2019). Available at https://www.greening.gov.hk/en/departments_greening_efforts/parks.html.

[65] Su, Alice, *op. cit.*

[66] Clennett, Britt and Marco Jakubec, "A lack of affordable housing feeds Hong Kong's discontent."

[67] Taylor, Chloe, "Hong Kong named world's most expensive city to buy a home" dated 12 April 2019 in CNBC (downloaded on 12 April 2019). Available at https://www.cnbc.com/2019/04/12/hong-kong-average-house-price-hits-1point2-million.html.

[68] Clennett, Britt and Marco Jakubec, *op. cit.*

[69] Jacobs, Harrison, "Could this be the solution to Hong Kong's housing crisis?" dated 22 May 2018 in Weforum (downloaded on 22 May 2018). Available at https://www.weforum.org/agenda/2018/05/the-most-expensive-city-in-the-world-may-build-100-square-foot-tube-homes-to-alleviate-its-escalating-housing-crisis [This article was published in collaboration with Business Insider].

[70] Lim, Janice, "Owning a home an 'impossible dream,' say Hong Kong youth frustrated over city's housing crisis."

[71] Jacobs, Harrison, *op. cit.*

Cramped conditions are common in Hong Kong. Many working adults and seniors in Hong Kong grew up in cramped spaces when they were young, using the same table for dinner, doing homework and so on. Even Hong Kong Chief Executive (CE) Carrie Lam was said to be raised in such an environment. Desperate to win over the trust and confidence of mainstream Hong Kong society, Lam is using housing policies to placate the complaints of high living costs in the territory. Even Chinese state-owned media noted the urgency of addressing the poor housing situation in Hong Kong. *China Daily* noted how Hong Kongers are compelled to live in shoe-box apartments that are less than 30 square metres, with bedrooms accommodating only a double bed and such spaces are getting smaller.[72]

About 200,000 Hong Kong residents stay in coffin homes that are so tiny that it is impossible to stretch or/and lift one's limbs.[73] According to a 2016 Hong Kong government report, a minimum of 209,700 Hong Kong residents stayed in subdivided apartments with an average of 57 square feet allocated to each individual, some of which are illegal, or have safety concerns.[74]

Disused or abandoned land and spaces in Hong Kong are rare. Former CE Carrie Lam used her annual address (Hong Kong's equivalent of the State of the Union address) to talk about her housing policies in very broad strokes, including using ordinances to resume undeveloped land in accordance with the law. The underdeveloped New Territories are the target of the proposed housing policies. Much of these lands are now storage areas, disused spaces and/or farmlands. The 1,500 hectares (approximately 2,545 acres) of agricultural and abandoned land are popularly known as brownfields for stationing vehicles and serving as shipping container parking lots and container yards.[75]

It may be a good use of land because some of them can never go back to traditional farming activities. Some of the rural land, formerly used for farming, were polluted by industrial activities and converted into parking

---

[72]Liang, Peter, "Mini-apartments help get people on housing ladder" dated 25 November 2015 in *China Daily* (downloaded on 1 January 2019). Available at http://www.chinadaily.com.cn/hkedition/2015-11/25/content_22516374.htm.

[73]Taylor, Alan, "The 'Coffin Homes' of Hong Kong" dated 16 May 2017 in The Atlantic (downloaded on 16 May 2017). Available at https://www.theatlantic.com/photo/2017/05/the-coffin-homes-of-hong-kong/526881/.

[74]Su, Alice, *op. cit.*

[75]Taylor, Alan, *op. cit.*

spaces, warehousing sites, junk depositories and recycling facilities.[76] Among such New Territories land, private land has been rezoned by the Hong Kong government for new town development like Hung Shui Kiu which has some 190 hectares of brownfield sites (permanently damaged agricultural land) and container storage areas.[77] As the land sits idle, there are voices urging the government to take it back for building public housing like the land in the New Territories.[78] Therefore, land reacquisition may start a precedent for other parts of Hong Kong too and hit other unexpected or unseen economic interests (some of which has government shares).

Among civil society groups, the environmentalists are in favour of using the New Territories' land for public housing. Andy Chu Kong, senior Greenpeace Hong Kong campaigner, opined that there are above 50 non-legal e-waste (electronic wastes) clearing sites, taking in toxic e-waste from other states and contaminating the environment as a result.[79] Hundreds of locations throughout the New Territories had effectively become the dumping ground for foreign electronic wastes, while the number of such locations has grown after mainland China banned imports of e-waste, 37 of such e-waste processing facilities that closed in the United States are making their way to Hong Kong.[80]

Forty percent of all containers that arrive in Hong Kong carry scraps (both e-wastes and non-e-wastes) while Hong Kong domestically generates more than 70,000 tons of e-wastes which are ironically exported overseas.[81] In 2015 alone, Hong Kong produced 3.7 million tonnes of waste and has already generated 13 landfill sites that are gentrified into green park spaces, golfing areas and physical activity areas and only three

[76] Su, Alice, *op. cit.*

[77] Brownlee, Ian, "Don't blame developers for Hong Kong's land 'shortage' – government policies are the real problem" dated 27 March 2018 in *South China Morning Post* (downloaded on 27 March 2018). Available at https://www.scmp.com/comment/insight-opinion/article/2139059/dont-blame-developers-hong-kongs-land-shortage-government.

[78] Bloomberg, "Hitting Tycoons Where It Hurts Could Appease Hong Kong Protesters."

[79] Su, Alice, *op. cit.*

[80] Standaert, Michael, "Welcome to Hong Kong, the world's dumping ground for electronic waste" dated 26 August 2017 in *South China Morning Post* (downloaded on 1 January 2019). Available at https://www.scmp.com/week-asia/society/article/2108339/welcome-hong-kong-worlds-dumping-ground-electronic-waste.

[81] Standaert, Michael, *op. cit.*

areas are left for landfill use.[82] Greenpeace Hong Kong advocates appear to be in favour of reusing the land as converting 50% of them to public apartment projects would come up to only US$33.3 billion (25% of the Lantau Tomorrow Vision project costs) and can supply 139,000 apartments.[83] The first phase was supposed to oversee the construction of manmade islands (approximately 1,000 hectares) next to Kau Yi Chau starting in 2025 with an eye to operate the initial housing spaces by 2032.[84]

The government's measures to take them back are known as land resumption policies and ordinances. The government can use their regulatory power as a disincentive for private development of the land because the approval process for developing the land is highly bureaucratic, time-consuming, expensive and so on. These disincentives can compel the landlords to embrace government projects and joint development.

To former Chief Executive Carrie Lam's credit, she is probably the most forthcoming and advocative CE in pushing forth measures to mitigate the housing crisis, a promise made during her election campaigning as well. In 2018, the then CE Lam came up with a vacancy tax to punitively target apartment hoarding and motivate property developers/owners to dispose of 1,600 apartments through quick sales within two months though the tax is not formally approved yet.[85]

The Lantau Tomorrow Vision: Former CE Lam spearheaded an initiative to construct 1,700 hectares (4200 acres) of manmade reclaimed land between Lantau and Hong Kong islands (the "Lantau Tomorrow Vision") in the SCS to fit 1.1 million Hong Kong residents by 2032 (1st phase).[86] But, it was rejected by critics for its US$63.8 billion (HK$500 billion)

---

[82]Robson, David, "Hong Kong has a monumental waste problem" dated 28 April 2017 in BBC (downloaded on 28 April 2017). Available at https://www.bbc.com/future/article/20170427-hong-kong-has-a-monumental-waste-problem.

[83]Su, Alice, *op. cit.*

[84]The Government of the Hong Kong Special Administrative Region, Hong Kong, "LCQ5: Lantau Tomorrow Vision" dated 14 November 2018 (downloaded on 14 November 2018). Available at https://www.info.gov.hk/gia/general/201811/14/P201811 1400683.htm.

[85]Su, Alice, *op. cit.*

[86]Clennett, Britt and Marco Jakubec, *op. cit.*

budget, 50% of the territory's fiscal reserves (US$140 billion or HK$1.1 trillion).[87]

The first phase of the vision requires 2,471 acres of land to create a business area and 260,000 apartments (70% public housing) connected by public transportation system, drawing from construction experiences gained from building Hong Kong Disneyland, Hong Kong International Airport (HKIA) and the 18.6 miles Macau-Hong Kong-Zhuhai bridge that took nine years to complete in 2018.[88] However, critics were also suspicious and speculated that the Lantau Tomorrow Vision project was to integrate Hong Kong with the Mainland, especially with the GBA project.

The Hong Kong authorities established the Steering Committee on Climate Change chaired by the chief secretary for administration in 2016 to consolidate the efforts of different local government bureaus on climate change and its challenges.[89] Believers of climate change like Professor Lam Chiu-ying (Department of Geography at the Chinese University of Hong Kong CUHK) warned of unpredictable conditions, such as super-typhoons, tidal storms, water surges and increase in sea level.[90]

The Hong Kong authorities claim they have already taken into account climate change issues when carrying out marine-based construction. For sea-based infrastructure constructions, the Hong Kong Civil Engineering and Development Department came up with the latest edition of the Port Works Design Manual in early 2018, with an eye on future climate change extrapolated in the Fifth Assessment Report of the United Nations Intergovernmental Panel on Climate Change.[91] Such manuals are meant to guide any future Hong Kong mega projects in the sea.

Without extracting marine mud, two possible construction techniques are: (1) inserting cement under the seabed mud and then erecting pillars for above-water level construction or (2) constructing a new manmade island created by a platform supported by gigantic solid steel structures stuffed with debris sitting on top of the seabed.[92] The Hong Kong government argued that to strengthen itself against bad climatic conditions, it

---

[87] *Ibid.*

[88] Su, Alice, *op. cit.*

[89] The Government of the Hong Kong Special Administrative Region, Hong Kong, *op. cit.*

[90] Su, Alice, *op. cit.*

[91] The Government of the Hong Kong Special Administrative Region, Hong Kong, *op. cit.*

[92] Su, Alice, *op. cit.*

intends to construct taller breakwaters, wave breakers and non-building buffer zones along the coastal areas.[93] Mainland Chinese environmental technologies and know-how may be potentially useful here if they were deployed by the HK government.

Innovative Solutions: Other innovative and unconventional solutions have also been offered such as James Law's utilization of sturdy retro-fitted concrete water pipes to build open-source O-Pods using more affordable Shenzhen's labour resources starting from April 2018.[94] The idea was inspired by his observation of remnant pipes left at construction sites. Standardized O-Pods are stackable and can fit into many urban spaces, including multistorey structures, office open park spaces and speedways.[95] In fact, pod housing is already quite developed in the West, with pod residents nicknamed "pod-estrians" and pod-sharing with an emphasis on affordable rental housing instead of ownership a common phenomenon.[96]

Aesthetically, the curved walls and angled lighting creates spacious feel and the pods come furnished with shelves, desk/dining table, wash-rooms and its modular structure can constantly evolve with needs (includ-ing turning them "smart" with Industry 4.0 tech).[97] All these were achieved for only US$15,000 and rentable at US$400 monthly (two-thirds invested in a savings account for the tenant and the rest for maintenance).[98] Simply put, it is intended to be operated like a social enterprise or non-profit outfit. It also has good prospects for export since many international companies are competing for a share of this global low-cost housing industry. Hong Kong's neighbour Shenzhen has become a major exporter

---

[93] The Government of the Hong Kong Special Administrative Region, Hong Kong, *op. cit.*

[94] Jacobs, Harrison, *op. cit.*

[95] Jacobs, Harrison, "The most expensive city in the world may build 100-square foot 'tube homes' to alleviate its escalating housing crisis" dated 21 May 2018 in Business Insider (downloaded on 21 May 2018). Available at https://www.businessinsider.com/hong-kong-housing-crisis-james-law-cybertectures-tube-homes-2018-5.

[96] Yuan, Callia, "PodShare offers affordable housing, sacrifices privacy of users" dated 22 September 2019 in the Stanford Daily (downloaded on 22 September 2019). Avail-able at https://www.stanforddaily.com/2019/09/22/podshare-offers-affordable-housing-sacrifices-privacy-of-users/.

[97] Jacobs, Harrison, *op. cit.*

[98] *Ibid.*

of pre-fabricated housing for the world and China's conservative media *Global Times* claims that its exports have brought down housing prices in New Zealand by 30%.[99]

Hong Kong's indigenously designed prototype, a 100 square feet O-Pod mock-up model displayed in Kwun Tong in East Kowloon, has the advantages of convenient maintenance and easy construction with mass-manufactured affordable quality that has good insulation/waterproofing/fireproofing properties and stackable without pillars, beams and columns (compared to low-cost container boxes).[100] It represents a possible option for low-cost manufactured housing for the masses.

Conclusion

The GBA region is placing high importance on tech such as 5G, IoTs, industrial internet, cloud computing, blockchain, data/smart computing centres and smart transportation, upgrades in tech infrastructure, data networks to support incubators/accelerators, better land use policies for releasing more affordable land, opening up domestic industries to foreign investors/competitors.[101] Professor Zhou Qiren of the National School of Development at Peking University (PKU) warned that China's economic advantage in the last 4 decades is slowly receding and China has not yet found a niche and persuaded Chinese firms to allocate more resources and efforts into tech innovation and capitalize on their latecomer advantage:

> Entrepreneurial spirit should be redefined…. In the past, efforts to lead people to affluence, to make good products, to fulfil social responsibility were called entrepreneurial spirit. Now and in the future, Chinese enterprises should focus more on enhancing the 'uniqueness' of technological innovation and strengthening late-mover advantage in innovation.[102]

---

[99] Wang, Cong and Zhang Hongpei, "Chinese-made prefabricated houses bring down housing costs for Kiwis by 30%" dated 10 July 2019 in the *Global Times* (downloaded on 10 July 2018). Available at http://www.globaltimes.cn/content/1110183.shtml.

[100] Jacobs, Harrison, *op. cit.*

[101] Ernst and Young (EY) Greater China, *op. cit.*

[102] Zhou, Mo, *op. cit.*

GBA innovation and technology growth is supported by larger numbers of Fortune 500 firms, the development of smart infrastructure and the Chinese state promoting Fintech, blockchain, big data and artificial intelligence (AI) to expand emerging industries.[103] The foreign investment law dated 1 January 2020 protects foreign investors from IPR infringements and obligatory technology transfers while state initiatives prioritize the creation of evaluation tools for technology-related intellectual property and developing insurance products to reduce the cost for tech enterprises.[104] Some argued there are new tech areas for GBA to enter into. Donald Li, a member/national political advisor of the 13th National Committee of the Chinese People's Political Consultative Conference (CPPCC) and President of the World Organization of Family Doctors, argued at the Asia Summit on Global Health in Hong Kong in November 2021 that GBA can tap into healthcare innovation (including community health and wellness).[105]

Li articulated that, because of the COVID-19 coronavirus pandemic, the healthcare industry has lured public interest, deals and investments, that can capitalize on GBA's plentiful healthcare resources/infrastructure for development: "The demand for healthcare services is growing in Asia rapidly, the provider networks in Asia are still developing, and the buildup of the healthcare ecosystem is likely to accelerate."[106] In December 2020, Hong Kong unfurled its "Smart City Blueprint 2.0" with 130 smart city initiatives related to mobility, healthcare, environmental protection, education, government services and other aspects and the blueprint also articulates how I&T can be utilized to mitigate COVID-19 coronavirus.[107]

Having discussed Hong Kong tech enterprises' role in developing their technological sector and the embeddedness of Hong Kong developments in the context of GBA regionalism, the following chapter focuses on specific sectoral case studies, including maritime technologies. It complements the historical chapter on China's pre-modern maritime technological development in the "Case Study: History of Pre-Modern Chinese Maritime Technologies" found in "Introduction."

---

[103] KPMG, *op. cit.*, p. 5.

[104] Ernst and Young (EY) Greater China, *op. cit.*

[105] Xinhua, *op. cit.*

[106] *Ibid.*

[107] KPMG, *op. cit.*, p. 3.

Bibliography

Bloomberg, "Hitting Tycoons Where It Hurts Could Appease Hong Kong Protesters" dated 22 August 2019 in Bloomberg (downloaded on 22 August 2019). Available at https://www.bloomberg.com/news/articles/2019-08-21/ hitting-tycoons-where-it-hurts-could-appease-hong-kong-protesters.

Brownlee, Ian, "Don't blame developers for Hong Kong's land 'shortage' – government policies are the real problem" dated 27 March 2018 in South China Morning Post (downloaded on 27 March 2018). Available at https:// www.scmp.com/comment/insight-opinion/article/2139059/dont-blame-developers-hong-kongs-land-shortage-government.

Clennett, Britt and Marco Jakubec, "A lack of affordable housing feeds Hong Kong's discontent" dated 11 August 2019 in Aljazeera (downloaded on 11 August 2019). Available at https://www.aljazeera.com/ajimpact/ lack-affordable-housing-feeds-hong-kong-discontent-190801151538867. html.

Constitutional and Mainland Affairs Bureau, "Innovation and Technology" dated 2018 in the Greater Bay Area website (downloaded on 1 January 2022). Available at https://www.bayarea.gov.hk/en/opportunities/it.html.

Ernst and Young (EY) Greater China, "Strategy for a tech-driven Greater Bay Area" dated 11 June 2021 in EY (Ernst and Young website) (downloaded on 11 June 2021). Available at https://www.ey.com/en_cn/china-opportunities/ strategy-for-a-tech-driven-greater-bay-area.

Greening, Landscape and Tree Management Section Development Bureau, "Parks, Country and Marine Parks" dated 2016 in the Greening, Landscape and Tree Management website (downloaded on 1 January 2019). Available at https://www.greening.gov.hk/en/departments_greening_efforts/parks. html.

Jacobs, Harrison, "Could this be the solution to Hong Kong's housing crisis?" dated 22 May 2018 in Weforum (downloaded on 22 May 2018). Available at https://www.weforum.org/agenda/2018/05/the-most-expensive-city-in-the-world-may-build-100-square-foot-tube-homes-to-alleviate-its-escalating-housing-crisis [This article was published in collaboration with Business Insider].

KPMG, "Greater Bay Area Update" in KPMG Tech and Innovation GBA newsletter dated February 2021 (downloaded on 1 March 2021). Available at https://assets.kpmg/content/dam/kpmg/cn/pdf/en/2021/02/greater-bay-area-update-tech-and-innovation-gba-newsletter.pdf.

Liang, Peter, "Mini-apartments help get people on housing ladder" dated 25 November 2015 in China Daily (downloaded on 1 January 2019). Available at http://www.chinadaily.com.cn/hkedition/2015-11/25/content_ 22516374.htm.

Lim, Janice, "Owning a home an 'impossible dream', say Hong Kong youth frustrated over city's housing crisis" dated 10 August 2019 in Today Online (downloaded on 15 November 2019). Available at https://www.todayonline. com/owning-home-impossible-dream-say-hong-kong-youth-frustrated-over-citys-housing-crisis.

Lim, Tai Wei, "Housing Policies in Hong Kong" dated 6 February 2020 in EAI Background Brief No. 1500 (Singapore: NUS EAI), 2020.

Robson, David, "Hong Kong has a monumental waste problem" dated 28 April 2017 in BBC (downloaded on 28 April 2017). Available at https://www.bbc. com/future/article/20170427-hong-kong-has-a-monumental-waste-problem.

Standaert, Michael, "Welcome to Hong Kong, the world's dumping ground for electronic waste" dated 26 August 2017 in South China Morning Post (downloaded on 1 January 2019). Available at https://www.scmp.com/week-asia/society/article/2108339/welcome-hong-kong-worlds-dumping-ground-electronic-waste.

Su, Alice, "Hong Kong aims to solve its housing crisis with an $80-billion artificial island" dated 3 April 2019 in Los Angeles Times (LA Times) (downloaded on 3 April 2019). Available at https://www.latimes.com/world/la-fg-hong-kong-housing-island-20190403-story.html.

Taylor, Alan, "The 'Coffin Homes' of Hong Kong" dated 16 May 2017 in The Atlantic (downloaded on 16 May 2017). Available at https://www.theatlantic. com/photo/2017/05/the-coffin-homes-of-hong-kong/526881/.

Taylor, Chloe, "Hong Kong named world's most expensive city to buy a home" dated 12 April 2019 in CNBC (downloaded on 12 April 2019). Available at https://www.cnbc.com/2019/04/12/hong-kong-average-house-price-hits-1point2-million.html.

Wong, Maggie, "Tai Mo Shan: How to tackle Hong Kong's highest peak" dated 7 February 2018 in CNN (downloaded on 7 February 2018). Available at https://edition.cnn.com/travel/article/tai-mo-shan-hong-kong/index.html.

Xinhua, "GBA offers huge potential for health tech investment" dated 24 November 2021 in China Daily (downloaded on 24 November 2021). Available at https://www.chinadailyhk.com/article/248942.

Zhou, Mo, "GBA seen as new engine driving China's economic growth" dated 22 September 2020 in China Daily (downloaded on 22 September 2020). Available at https://www.chinadailyhk.com/article/144097#GBA-seen-as-new-engine-driving-China's-economic-growth.

Chapter 9

Hong Kong's Maritime Technologies[1]

This chapter on the first case study of "Hong Kong's Maritime Technologies" details the development of Hong Kong as a global port city with sophisticated container traffic and shipping activities, including the roles played by its shipping companies and public policy.

Introduction

In Hong Kong's golden era of container logistical business in 2004, Hong Kong's container port throughput came up to 21.98 million TEUs, becoming the globe's busiest container port then.[2] At the end of 2019, Hong Kong was internationally the 4th biggest shipping register after Panama, Liberia and Marshall Islands with registered vessels making up 130 million gross tonnes (GT, December 2020 data).[3]

With over 1.5 centuries of maritime heritage and strong global connectivity, HKSAR is within the top 10 ranking container ports globally, a transhipment hub and approximately 280 weekly container ship traffic

---

[1] Drawn from a small portion of the author's manuscript on Maritime Silk Road.

[2] Poon, C. H., "Hong Kong Logistics: Maritime Prospects and Trends" dated 12 October 2021 in HKTDC (Hong Kong Trade and Development Council) Research (downloaded on 12 October 2021). Available at https://research.hktdc.com/en/article/ODc2MTE1OTM1.

[3] Fung, Doris, "Maritime Services Industry in Hong Kong" dated 19 March 2021 in Hong Kong Trade Development Council (HKTDC) (downloaded on 19 March 2021). Available at https://research.hktdc.com/en/article/MzExMzA4Mzc2.

linked to more than 600 destinations globally.[4] Hong Kong was the globe's 8th biggest trading economy in 2019 with global trade greatly assisted by its efficient port, something acknowledged by ship owners, cargo owners and traders even as the port city enlarge its maritime service cluster offering ship management, ship broking, ship finance, maritime insurance and legal services.[5]

Technology is an important factor in driving Hong Kong's ranking position in port operations and activities, making it one of the global leaders in this area. The ship planning systems, for example, optimize the offloads and uploads on ships, efficient container management, ship storage and yard planning through high-tech computer systems to facilitate effective container yard procedures.[6] Hong Kong's technological and technical expertise extends even into the internal waterways of China. As early as 1998, to satisfy the accelerated increase in passenger volume between Hong Kong and the Pearl River Delta, Hong Kong constructed the globe's biggest fleet of high-speed ferries along with substantial experience/expertise to maintain and managed these ferries.[7] Internal waterways' technological infrastructure upgrades are also mentioned in Beijing's blueprints. In the Outline Development Plan for the Guangdong-Hong Kong-Macao GBA, Beijing is ensuring building more capacity of global shipping services in Guangzhou/Shenzhen and service capacity of infrastructural facilities such as ports and fairways, creating a coordinated system of ports, shipping, logistics and ancillary services with Hong Kong and improving cargo collection/distribution in inland waterways/port railway lines.[8]

---

[4]Onag, Gigi, "HK maritime industry urged to be green and smart" dated 11 November 2021 in Futureiot (downloaded on 11 November 2021). Available at https://futureiot.tech/hk-maritime-industry-urged-to-be-green-and-smart/.

[5]Fung, Doris, *op. cit.*

[6]Hong Kong Container Terminal Operators Association (HKCTOA), "Leading Edge Technology" dated 2022 in Hong Kong Container Terminal Operators Association (HKCTOA) website (downloaded on 6 February 2022). Available at http://www.hkctoa.com/technology.

[7]Ip, Stephen, "Hong Kong as an international shipping centre" dated 25 June 1998 in Hong Kong Government Info Daily Information Bulletin website (downloaded 1 January 2020). Available at https://www.info.gov.hk/gia/general/199806/25/0625089.htm.

[8]Poon, C. H., *op. cit.*

And, as early as 1998, then Secretary for Economic Services Stephen Ip stressed at the second "Conference for New Ship and Marine Technology into the 21st Century" (an international meeting of maritime field specialists) that Hong Kong remains active in the shipping industry even though by then, the SAR has been transiting to the financial and services sector.[9] In 1998, port and shipping activities made up 20% of HK's GDP and HK ship owners manage some 6% of the global tonnage (31 million gross tonnes) and operate a shipping register of 500 ships over six million gross tonnes, with 1,000 shipping services firms in ship management, brokerage, financing and insurance and arbitration.[10]

In terms of technological implementation, Hong Kong container terminal operators utilized Electronic Data Interchange (EDI) since 1988 to facilitate effective information flow between shipping firms and terminal operators.[11] Some of its ports underwent tech transformation in the 2010s. For example, The Mawan Smart Port at Shekou, Shenzhen, came into being in late June 2021 against the context of three decades of operations of the Port of Mawan itself, had transformed its manpower-intensive bulk cargo business into a high-tech port starting from 2017 through collaboration with ranking internet technology (IT) firms.[12]

Fast-forwarding ahead to the contemporary era, Hong Kong Shipping Register (HKSR) (differentiated from China) has listed approximately 2,600 vessels (December 2020 data) 130 million GT and is the globe's 4th biggest ship register after Panama, Liberia and Marshall Islands by the end of 2019.[13] Many are attracted to Hong Kong also for its taxation reasons. Income from international trade carried out by HK-registered ships is exempted from profits tax and Hong Kong has double taxation agreements (DTAs that include cover shipping income) with the US, the UK, the Netherlands, Denmark, Norway and Germany, making it a low tax regime.[14]

---

[9] Ip, Stephen, *op. cit.*

[10] *Ibid.*

[11] Hong Kong Container Terminal Operators Association (HKCTOA), *op. cit.*

[12] Poon, C. H., *op. cit.*

[13] Fung, Doris, *op. cit.*

[14] *Ibid.*

Maritime Services

On 1 June 1998, to dispense and organize a spectrum of services and activities, the Hong Kong government founded the Hong Kong Port and Maritime Board to coordinate state/private sector efforts to market Hong Kong as a global shipping hub, provide support in technological fields such as telecommunications to lure more Chinese/foreign shipping firms to HK. Consequently, today, Hong Kong's maritime service cluster provides comprehensive professional services, including ship financing, insurance and broking, ship management and maritime legal advice.[15]

Hong Kong has a long tradition of organization important international conferences in the past. For example, the Hong Kong Institute of Marine Technology, the Hong Kong Joint Branch of the Royal Institute of Naval Architects, the Institute of Marine Engineers and the Hong Kong Institute of Engineers organized the "Conference for New Ship and Marine Technology into the 21st Century" in 1998 for the first time.[16] The yearly ALMAC held by the HK Government and the Hong Kong Trade Development Council is a flagship Hong Kong Maritime Week activity that showcases physical/virtual exhibitions featuring logistical technologies, 5G warehouse management tech, smart logistics solutions, international payment solutions and smart port developments while linking up participants with optimal business solutions.[17]

An example of such smart systems can be found in the Kwai Tsing Port. Kwai Tsing Port Gate Automation is implemented through Information Exchange Services, tractor booking systems and gatehouse operations with its paperless common tractor identity smart cards (TID) for external tractors to bypass the requirement for five different cards to be dispensed to each tractor driver for entry into the five different terminals, substantially reducing time wasted in the process.[18] Another example is in Mawan. The Mawan Smart Port has in place smart features like the China Merchants Core smart business platform, China Merchants one-stop online customer service platform ePort, artificial intelligence (AI),

---

[15] *Ibid.*

[16] Ip, Stephen, *op. cit.*

[17] Onag, Gigi, *op. cit.*

[18] Hong Kong Container Terminal Operators Association (HKCTOA), *op. cit.*

5G applications, BeiDou Navigation Satellite System (BDNSS), automation, smart customs, blockchain and green low-carbon operation.[19]

In her presentation at the 11th Asian Logistics, Maritime & Aviation Conference (ALMAC) in Hong Kong in November 2021, Lam's administration is keen to utilize innovative technologies to augment maritime services in ship financing, marine insurance, maritime legal/arbitration services, ship agency management/shipbroking by putting in place economic incentives like tax concessions and manpower training.[20] The Hong Kong University of Science and Technology is one of the established educational/research hubs for technical studies in East Asia with notable reputed engineering courses and research activities, contributing to the HK marine training industry churning out naval architects, marine engineers and surveyors.[21] As a base to large numbers of longstanding professional ship management services providers, some provide consultancy services in ship engineering, construction and shipyard selection.[22]

Another university active in marine environment research is Hong Kong University. Established in 2020 by marine biology professor David Baker and PhD student Vriko Yu, their firm archiREEF has been conserving coral at Hoi Ha Wan Marine Park coastal reserve since 2016, in collaboration with the Agriculture, Fisheries and Conservation Department (AFCD) and Hong Kong University's architecture department by developing an artificial coral reef using the HKU's 3D printing facility.[23] Such technologies can help reverse pollution caused by maritime activities.

Logistics

Aside from infrastructural facilities, supply chain management (SCM) requires professional services which fit Hong Kong's optimal regional location and mature marine-related high value-added supply chain services, prompting global shipping and maritime service firms to utilize

---

[19] Poon, C. H., *op. cit.*

[20] Onag, Gigi, *op. cit.*

[21] Ip, Stephen, *op. cit.*

[22] Fung, Doris, *op. cit.*

[23] Cairns, Rebecca, "How 3D printing could help save Hong Kong's coral" dated 13 September 2021 in CNN Travel (downloaded on 13 September 2021). Available at https://edition.cnn.com/travel/article/hong-kong-3d-printing-coral-reefs-hnk-spc-intl/index.html.

Hong Kong as their platform in this areas.[24] At the 11th ALMAC in Hong Kong in November 2021, Hong Kong's Secretary for transport and housing Frank Chan highlighted the significance of importance innovative technologies to stay ahead in global logistics, maritime and aviation hub:

> The future of modern logistics is going to be smart and technology-driven. Automation, AI, big data and digitalisation are instrumental, said Chan, adding that to help achieve this strategic goal, the Hong Kong government has up a HK$300 million funding scheme to encourage logistics service providers in applying technological solutions to enhance productivity.[25]

Hong Kong may be keen to implement advanced digital technologies. They include smart port technology and become the demonstrative platform for R&D, implementation of smart waterways in Guangdong, Hong Kong and Macao (also spelt Macau), utilization of BDNSS and the integrated implementation of IoT to shore up its comparative advantages in the international logistics supply chain.[26] Hong Kong's Kwai Tsing Port, for example, manages millions of containers by utilizing their separate terminal management systems that integrate and optimize gate/yard/ship operations, optimized operational parameters, a common database for terminal activities, real-time data updating and monitoring system.[27]

According to World Bank's 2018 Logistics Performance Index (LPI), Hong Kong chalked up 3.92 points to rank 12th globally (3rd in Asia), 7th out of 160 economies in global shipment performance, 8th biggest trading economy in 2019, a global maritime centre with a large community of ship owners/cargo owners/traders situated in a robust maritime services cluster.[28] The implementation of technologies in the logistical sector is also not new. Given that container cargo is distributed from a comprehensive production base in southern China, a Barge Centre was

---

[24]Poon, C. H., *op. cit.*

[25]Onag, Gigi, *op. cit.*

[26]Poon, C. H., *op. cit.*

[27]Hong Kong Container Terminal Operators Association (HKCTOA), *op. cit.*

[28]Fung, Doris, *op. cit.*

built in 1998 with technologies like a Barge Identity Card system in place to automate the identity authentication procedure in the context of growing business demand.[29]

Challenges: Because of Hong Kong's global transit hub, flexible policies, free port status, ranking trade and customs facilitation placement and reliability of transportation/logistics infrastructure/services (all of which surpassed the indicators for its mainland counterparts), many multinational firms selected Hong Kong for inventory management, labelling, packaging and other processing procedures.[30] However, its competitors, both in China and overseas, are catching up. Hong Kong has mature ports and logistics infrastructure within China and is encountering changes in the manufacturing sector in the Pearl River Delta (PRD) and so it needs to tap into technologies to handle challenges and enhance efficiencies.[31] Dr. Robin Li Yubin, Vice President (VP) of China Merchants Port Holdings Co. Ltd., urged Hong Kong to sustain strategic collaboration with partners in IT to attain the strategic intelligent digital transformation of port enterprises and energize port logistics businesses.[32] Mainland Chinese ports have become leaner and greener. For example, the Guang'ao Port Area Shantou Port's shore-side electricity project has enable berthing vessels to turn off their generators and charge with pier-side equipment, minimizing atmospheric emissions.[33]

Hong Kong's business leaders appear to agree with the view to collaborate more with Chinese companies. Victor Mok, chairman and CEO of Asset Service Platform at GLP China noted that Chinese logistics firms have utilized digital solutions 2011–2016 and beyond to augment the digital transparency and cargo logistical safety and speed up cross-boundary transnational e-commerce.[34] After all, China's master plan promises to support Hong Kong's tech development in the port industry. In the outline of the 14th FYP for National Economic and Social Development and the Long-Range Objectives Through the Year 2035, China has pledged to

---

[29] Hong Kong Container Terminal Operators Association (HKCTOA), *op. cit.*

[30] Poon, C. H., *op. cit.*

[31] Fung, Doris, *op. cit.*

[32] Poon, C. H., *op. cit.*

[33] *Ibid.*

[34] Onag, Gigi, *op. cit.*

accelerate world-class port clustering in the GBA, expand clean energy, ecological environment and infrastructure upgrades and speed up construction of intercity railways.[35]

Hong Kong is determined to keep up to speed with the technological progress that mainland and foreign rivals are making. Some of the newest techs can be seen in facilities like the Kwai Tsing Port. Kwai Tsing Port operates an effective and dependable container port with coordinated functional systems to prevent inefficiency and duplication of processes for port end users, including the Terminal Management System, computer screen, gate automation and EDI Technology.[36] HIT is the first container terminal operator in Hong Kong to use remote-controlled rubber-tyre gantry cranes (RTGCs), automated container stacking system, surveillance cameras and precision sensors in the 29 cranes at Kwai Tsing Port, all of which improved efficiency and operational safety.[37]

Raymond Fung, director of trades at Orient Overseas Container Line advocated using e-commerce and artificial intelligence (A.I.) to project/forecast demand for shipping firms/lines to tailor ships of different sizes and routings to meet customers' needs.[38] For example, the Mawan smart port is 983,600 square metres in terms of size with a berth shoreline of 1930 metres and five berths, including two 200,000-tonne container berths for the largest liners, making it the biggest in southern China and able to manage 3 million TEUs yearly.[39] Other than the facilities, the shippers that use them have also upgraded their systems. Mark Slade, managing director of DHL Global Forwarding Hong Kong & Macau, explained how DHL improve their logistical services: "We at DHL use an analytical tool to allow companies to get deeper into the supply chain and identify risks with suppliers that are actually two or three layers removed from their operations."[40]

COVID-19 coronavirus pandemic, Brexit, the Suez Canal blockage and other external shocks have plunged the international container shipping market into difficulties in 2021, resulting in supply-demand misalignment, container insufficiency and port crowdedness and their ripple

---

[35] Poon, C. H., *op. cit.*

[36] Hong Kong Container Terminal Operators Association (HKCTOA), *op. cit.*

[37] Fung, Doris, *op. cit.*

[38] Onag, Gigi, *op. cit.*

[39] Poon, C. H., *op. cit.*

[40] Onag, Gigi, *op. cit.*

effects on freight rates/shipment reliability but as the pandemic becomes an endemic, normality may return and freight rates may taper off.[41] During the COVID-19 coronavirus pandemic, supply chain disruptions hurt the air cargo and logistics industry, so many companies are using technologies to secure supply chains even before the start of the pandemic.[42] It will take the effort of all shippers to practice epidemic prevention and mitigation to facilitate effective maritime logistical channels and safe/steady supply chain activities, and vaccinations to eventually mitigate the COVID-19 coronavirus pandemic so that global economic/trading recovery can take place for sustainable development of port logistics.[43]

Concluding Remarks

While in 2020, HK managed to handle 17.97 million TEUs (9th globally), Hong Kong's container throughput is still ranked top 10 in the world despite strong competitive pressure from proximate ports and so Dr. Robin Li Yubin, Vice President (VP) China Merchants Port Holdings Co. Ltd, believes that it is due to "sound logistics infrastructural support and professional logistics service capability."[44] According to the International Shipping Centre Development (ISCD) Index 2020 published by the Baltic Exchange and Xinhua (China's state-owned media outfit), Hong Kong's ranking as a global maritime centre is only behind Singapore, London and Shanghai while it was the 9th busiest container port in the world in 2020 after Shanghai, Singapore, Ningbo-Zhoushan, Shenzhen, Guangzhou, Qingdao, Busan and Tianjin.[45] But, it has mature infrastructure that needs to catch up technologically.

The international port logistics industry is facing the need for digitalization and intelligent transformation and its effects on supply chain management and so firms need to share digitalization and intelligent transformation, accelerated by the COVID-19 coronavirus pandemic. Here, Hong Kong can look for mainland Chinese partners for tech upgrades. China's Guangzhou Port Group constructed the first 5G Nansha

---

[41] Poon, C. H., *op. cit.*

[42] Onag, Gigi, *op. cit.*

[43] Poon, C. H., *op. cit.*

[44] Hong Kong Container Terminal Operators Association (HKCTOA), *op. cit.*

[45] Fung, Doris, *op. cit.*

smart port in a team-up with Huawei, featuring next-gen IoT (Internet of Thing) sensors/AI, smart guided vehicles and automated container cranes and is the pioneering fully automated GBA terminal.[46]

Environmentalism: Threatened by climate change, many are unaware that Hong Kong has one of Asia's most biodiverse hard coral species (more than the Caribbean) but is under threat from development and pollution, so one of its companies is turning to 3D-printed terracotta tiles to help corals grow and save ocean life.[47] This is just one of an ecological system of efforts to restore Hong Kong's marine environment. Hong Kong's port system is keen to become carbon neutral and adopt a carbon-peaking strategy by putting in place green innovations to construct green ports' supply chains by giving preferences for green equipment, augmenting climate change-related risks management, organizing online environmental protection activities and practice energy conservation/emissions reduction concepts in everyday scenarios.[48] Hong Kong company archiREEF's green innovation has capture global attention. The company makes two feet tiles to imitate the platygyra brain coral to lure marine life hiding from predators and baby corals are attached to the tile with non-toxic glue for them to grow on, in turn attracting marine life to live in such reefs which, when mature, would erode the tile foundation.[49]

The Hong Kong authorities are also keen to switch the maritime industry to utilize clean energy and enhance efficiency/productivity through smart port strategies as Hong Kong Chief Executive (HKCE) Carrie Lam articulated the following:

> While we seek to expand our maritime industry, we have not forgotten our commitment to sustainable development....We have already announced the target for Hong Kong to achieve carbon neutrality before 2050. As part of our decarbonisation effort, we encourage industry players to adopt more sustainable shipping initiatives. As you all know, Hong Kong was the first city in Asia to mandate a fuel switching requirement for ocean-going vessels.[50]

---

[46] *Ibid.*

[47] Cairns, Rebecca, *op. cit.*

[48] Poon, C. H., *op. cit.*

[49] Cairns, Rebecca, *op. cit.*

[50] Onag, Gigi, *op. cit.*

The core port business implemented new energy conservation technologies/products. They include the following: "fuel to electricity conversion" and "shore-based power supply for ships" and from 2020 to 2021, a fuel-to-electricity conversion project for Shantou Port's rubber-tyred container gantry cranes, replacing fuel-powered with electricity-powered equipment, utilizing clean low-carbon electricity instead of fuel, all with the objectives of zero exhaust emissions and increasing energy utilization rates.[51]

Hong Kong has the potential to export its green systems overseas. Through "innovation, coordination, green, openness and sharing" in all business activities, Hong Kong projects built an energy-conserving, low-carbon port system such as LED lights, rubber-tyred container gantry cranes, non-diesel advanced electric lifters at the Sri Lankan Colombo International Container Terminal (CICT) to cut down carbon emissions by 6502 tonnes as the first green pier in Sri Lanka. Corporate and state customers sponsor archiREEF's conservation efforts to manage sites for up to 5 years tracking its improved biodiversity and their use of non-poisonous nature clay supersedes all current designs used by other leading companies like Australia's Reef Design Lab and D-Shape.[52] For large corporate and state clients (perhaps in HK or otherwise), archiREEF offers tailorable 3D printing technology for the corals in different conditions and environmental challenges to regenerate and maintain Earth's underwater ecosystems which are valuable carbon stores.[53]

Finally, due to China's "dual circulation" policy, Hong Kong is firmly embedded in the development of the Guangdong-Hong Kong-Macao GBA and will likely prosper from such regional development.[54] In Beijing's 14th FYP, the authorities will support Hong Kong in augmenting its shipping and trade centre status for dual domestic and global circulation, including Hong Kong's efforts in digitalization, automation and smart tech implementation in both integrated upstream/downstream logistical supply chains and niched supply chain services.[55] In the *Outline Development Plan for the Guangdong-Hong Kong-Macao Greater Bay*

---

[51] Poon, C. H., *op. cit.*

[52] Cairns, Rebecca, *op. cit.*

[53] *Ibid.*

[54] Hong Kong Container Terminal Operators Association (HKCTOA), *op. cit.*

[55] Poon, C. H., *op. cit.*

Area, Beijing has started the objectives of building modern freight/logistics systems, provision of railway-water, motorway-railway, air-railway and river-sea inter-modal transport and implementing "single cargo manifest" services and has put in place a smart transport system with transport IT, IoT, cloud computing, big data, etc.[56]

Dr. Robin Li Yubin, Vice President (VP) of China Merchants Port Holdings Co. Ltd. urged regional cooperation and technological implementation as the future of downstream development and capture GBA opportunities by "extend its strengths and experience in shipping, port and logistics services to other cities in the GBA and expand the market to the whole world through the GBA."[57] For example, Hong Kong has the potential to export its green systems overseas. Through "innovation, coordination, green, openness and sharing" in all business activities, Hong Kong projects built an energy-conserving, low-carbon port system such as LED lights, rubber-tyred container gantry cranes, non-diesel advanced electric lifters at the Sri Lankan CICT to cut down carbon emissions by 6,502 tonnes as the first green pier in Sri Lanka.[58]

Finally, bringing Hong Kong's technological development up to speed with contemporary events, the final case study in the Hong Kong section of this volume focuses on the technologies that Hong Kong used to mitigate COVID-19 coronavirus pandemic. This is discussed in the following chapter by Elim Wong Yee Lam.

Bibliography

Cairns, Rebecca, "How 3D printing could help save Hong Kong's coral" dated 13 September 2021 in CNN Travel (downloaded on 13 September 2021). Available at https://edition.cnn.com/travel/article/hong-kong-3d-printing-coral-reefs-hnk-spc-intl/index.html.

Fung, Doris, "Maritime Services Industry in Hong Kong" dated 19 March 2021 in Hong Kong Trade Development Council (HKTDC) (downloaded on 19 March 2021). Available at https://research.hktdc.com/en/article/MzEx MzA4Mzc2.

Hong Kong Container Terminal Operators Association (HKCTOA), "Leading Edge Technology" dated 2022 in Hong Kong Container Terminal Operators

---

[56] *Ibid.*
[57] *Ibid.*
[58] *Ibid.*

Association (HKCTOA) website (downloaded on 6 February 2022). Available at http://www.hkctoa.com/technology.

Ip, Stephen, "Hong Kong as an international shipping centre" dated 25 June 1998 in Hong Kong Government Info Daily Information Bulletin website (downloaded on 1 January 2020). Available at https://www.info.gov.hk/gia/general/199806/25/0625089.htm.

Onag, Gigi, "HK maritime industry urged to be green and smart" dated 11 November 2021 in Futureiot (downloaded on 11 November 2021). Available at https://futureiot.tech/hk-maritime-industry-urged-to-be-green-and-smart/.

Poon, C. H., "Hong Kong Logistics: Maritime Prospects and Trends" dated 12 October 2021 in HKTDC (Hong Kong Trade and Development Council) Research (downloaded on 12 October 2021). Available at https://research.hktdc.com/en/article/ODc2MTE1OTM1.

Chapter 10

Utilization of Advanced Technology During the COVID-19 Pandemic in Hong Kong

Introduction

"Together, We Fight the Virus!" is the slogan announced by the Hong Kong Government in 2021. To fight the COVID-19 virus, Hong Kong citizens and the Hong Kong Government have been working together via the use of high technology: LeaveHomeSafe. LeaveHomeSafe is a mobile phone contact tracing application launched on 16 November 2020 by the Hong Kong Government. Since then, Hong Kong citizens started to live their lives with this technology whenever they go on the island: to the mall, to work, to dine or to visit public places. Massive use of mobile technology in Hong Kong is a new phenomenon. This chapter introduces the use of this COVID-19 tracing app in Hong Kong and probes the question of in what ways does it help to cut the spreading chain of COVID-19 in Hong Kong. Does this app successfully prevent Hong Kong from the sixth outbreak? Or how can high technology apply to daily life among Hong Kong citizens?

The Origins of LeaveHomeSafe

The first few Wuhan-related COVID-19 cases in Hong Kong were reported on 7 January 2020, but the use of high technology to trace the cases did not happen until November 2020, which is 10 months since the first wave of COVID-19 outbreak. The Free LeaveHomeSafe mobile application is a technology chosen by the Hong Kong Government for fighting the virus. This app is developed by the Office of the Government Chief Information Officer (OGCIO) and used to contact tracing and surveillance that aimed to trace COVID-19 spreaders. According to the Hong Kong Government, LeaveHomeSafe is "tapping technology to combat the pandemic, the app aims to encourage the public to keep a more precise record of their whereabouts minimizing the risk of further transmission of the virus and protect Hong Kong together."[1] When the mobile application was first made available for public download on 16 November 2020, it was not a compulsory measure. The government encouraged the public to use LeaveHomeSafe as a platform to record visits at users' discretion, but it is up to the individuals to choose to participate voluntarily.

To develop the first tracing app in Hong Kong history, the Hong Kong Government has invited over 6,000 public and private venues, including government office buildings, sports centres, swimming pools, libraries, markets, cooked food markets, community halls, shopping centres of public housing estates, hospitals, clinics, post office, construction sites, restaurants, bars or pubs, karaoke establishments, clubs, fitness centres and banks, to support the campaign. All the venue owners need to do is to put a Quick Response Code (QR Code) at the entrance and ask the visitors to scan the Code on the LeaveHomeSafe app. The app will then record the location and time of visit for the users. When visitors leave the venue, they will have to click "leave now" on the app and the time of departure will be recorded. Besides, LeaveHomeSafe also helps record each taxi trip for the passengers by scanning the registration mark located inside the taxi. By using LeaveHomeSafe, users will be notified when they happened to have visited the same venue as the COVID-19 patients at around the same time and remind them to do COVID-19 test.

---

[1] "Launch of "LeaveHomeSafe" COVID-19 exposure notification mobile app," The Government of the Hong Kong. Accessed 24 June 2022, https://www.info.gov.hk/gia/general/202011/11/P2020111100367.htm?fontSize=1.

When the mobile application was first announced in November 2020, the public showed concerns with regard to personal data privacy. They do not feel comfortable being tracked and having their whereabouts disclosed. The government has the following explanation for data protection:

> The "LeaveHomeSafe" mobile app upholds the principle of protecting personal data privacy, and user registration is not required. The app will not use positioning services or any other data of the users' mobile phones. Venue check-in data will be encrypted and saved on users' devices only. Such data will not be uploaded to the Government or any other systems. Check-in data will be kept in users' mobile phones for 31 days and will then be erased automatically.[2]

From the first day of release on 16 November 2020 to 15 December 2020, there were over 370,000 downloads of LeaveHomeSafe on app stores.[3] And since January 2022, LeaveHomeSafe is also an app that allows users to register Hong Kong citizens' Health Code on the app.

LeaveHomeSafe vs Handwritten Memo

Since using LeaveHomeSafe is based on voluntary participation, the Hong Kong Government provided another method to collect personal information in order to be traceable for public notification for COVID-19 confirmed cases — which is collecting handwritten personal information. The Hong Kong Government announced that all visitors who visit the selected venues where the app is in place, including restaurants, fitness centres, entertainment venues and beauty centres, need to use LeaveHomeSafe to check into the venue, or to keep the handwritten personal information for the venues to retain for 31 days. Personal data collected from handwritten memos include names and contact numbers of the customers. If there is COVID-19 confirmed case, the venue officers have the responsibility to contact all visitors who visited the venue at about the

---

[2] *Ibid.*

[3] "LCQ8: "LeaveHomeSafe" mobile application," The Government of the Hong Kong. Accessed 24 June 2022, https://www.info.gov.hk/gia/general/202012/16/P2020121600222. htm?fontSize=1.

same time on the same date. This measure of allowing visitors to fill in personal information on handwritten memos instead of using advanced technology helps to relieve the pressure on those who cannot afford smartphone or who are not familiar with high technology.

Why do people choose to fill in personal data on a piece of paper instead of using high technology, which is easy to use with just one click on the mobile application? *Citizen News* has conducted interviews with visitors who chose to hand in handwritten memos to the restaurant instead of using LeaveHomeSafe. In one of these interviews, Mr. Tang claimed that he does not want to use LeaveHomeSafe because he felt insecure and worried if his personal information would be stored or used elsewhere on the Internet.[4] Another interviewee Ms. Yung, on the other hand, explained that she usually fills in the handwritten memo in restaurants, but use LeaveHomeSafe for public hospital visit (which is the only way to enter the government buildings) with her second mobile phone.[5] The practice of using a second mobile phone becomes common because LeaveHomeSafe application does not require an Internet connection and people can install the application and use it on a non-daily use mobile phone.

There is no perfect measure to ensure the tracing of all visitors is accurate, and the most common failure of using handwritten memos as COVID-19 case trace is the invalid information provided by visitors. Since February 2021, when handwritten memos for personal data became compulsory for people who enter the selected venues, the Hong Kong Government has enforced measures to ensure the process goes smoothly, including regular inspection at the venues and fining those who failed to provide valid information or venue owners for not checking the information carefully. For example, on 15 October 2021, the Hong Kong Police Force reported that, during regular inspections carried out at restaurants, they found two customers who provided invalid contact numbers and one who did use LeaveHomeSafe or submitted handwritten memos to the

---

[4]"Handwritten-memo-visitors use LeaveHomeSafe for the first time, feel safer to install on the extra mobile phone," CitizenNews. Accessed 24 June 2022, https://www.youtube.com/watch?v=gqF0m3LWPQY.

[5]*Ibid.*

restaurant.[6] The restaurant owner was fined HK$8,000 for failing to ensure accurate tracking of customers.[7]

In fact, not only do the visitors who choose not to use LeaveHomeSafe have the responsibility to provide valid information for the use of COVID-19 case tracking, but the venues are also required to protect the customers' personal data and avoid dishonest use of the contacts. The Office of the Privacy Commissioner for Personal Data (PCPD) received 49 cases (31 cases on restaurants and 18 cases on other venues) on the dishonest use of personal data collected by the listed venues from handwritten memos from November 2020 to November 2021.[8] Among the 49 cases, 15 cases were charged, and the PCPD issued warning notices to those venues to make sure they will improve the way of collecting personal data from visitors.[9]

Lacking proper guidelines in personal information collection is often a problematic issue happened in Hong Kong as well. The Food and Environmental Hygiene Department of Hong Kong has issued a template for all venues to collect customer information records.[10] The template has clearly listed that a fine of HK$5,000 applies to all customers who provide invalid information under the Cap. 559F. However, this form is a suggested template and not compulsory for the venues to use. Therefore, there were a few disputes between the venues and customers over the way of collecting

---

[6]"Customers failed to use LeaveHomeSafe and filled in wrong personal data, Mongkok Hot pot restaurant owner is fined HK$8,000," Bastille Post. Accessed 24 June 2022, https://www.bastillepost.com/hongkong/article/9639102-%E9%A3%9F%E5%AE%A2%E7%84%A1%E7%94%A8%E5%AE%89%E5%BF%83%E5%87%BA%E8%A1%8C%E5%8F%8A%E4%BA%82%E5%A1%AB%E8%B3%87%E6%96%99-%E6%97%BA%E8%A7%92%E7%81%AB%E9%8D%8B%E5%BA%97%E8%B2%A0%E8%B2%AC%E4%BA%BA%E5%88%A4.

[7]*Ibid.*

[8]"Office of the Privacy Commissioner for Personal Data received 49 complain cases on handwritten memo, 15 cases are found, LeaveHomeSafe is better," CABLE TV & CABLE News. Accessed 24 June 2022, https://www.youtube.com/watch?v=bqMRv9oAIq8.

[9]"Office of the Privacy Commissioner received 49 cases in handwritten memo, 15 cases on privacy leak are found," Radio Television Hong Kong. Accessed 24 June 2022, https://news.rthk.hk/rthk/ch/component/k2/1621230-20211124.htm.

[10]"Customer information record form," Food and Environmental Hygiene Department. Accessed 24 June 2022, https://www.fehd.gov.hk/tc_chi/events/covid19/images/vaccine-Bubble_customer_record_form.pdf.

personal data. A Facebook user posted an article on Facebook to describe how her personal information was not treated properly by the restaurant she visited. After she wrote her name and contact number on the paper provided by the restaurant, she discovered the waiter passed the paper to another customer to fill in without covering up her personal information.[11] When she asked the waiter to hide her information before passing the paper to the next customer, the waiter refused, claimed that there is no such regulation to do so and blamed the customer for not using LeaveHomeSafe.[12]

Not only indicated on online platforms but similar issues were reported in newspapers too. *WenWeiPo* has published a feature on the disadvantages of using handwritten memos as a measure to trace COVID-19 cases. The article, published in November 2021, is a field observation piece. The *WenWeiPo* reporter visited more than twenty restaurants, and less than 10% of the restaurants used the template suggested by the Food and Environmental Hygiene Department.[13] The rest of the "health decoration form" does not include the content of Cap. 599F or require the customer to sign the agreement. The "form" is simply a paper with the name and contact number written on it. Another disadvantage of using handwritten memos as COVID-19 case trace is the difficulty of recordkeeping. As each venue collected a large number of handwritten memos every day, it is difficult to ensure all venues keep their daily records for 31 days. According to the news article, an anonymous worker in F&B industry confessed to *WenWeiPo* reporter that, most of the time, the memos were thrown away every night: "[we] received over hundreds of memos every day. And each of them needs to be kept for 31 days? So that would be thousands of memos? Where shall we store them [the memos]? No government office will come and check. They never did."[14]

---

[11] "Customer discovered personal data is disclose in public after placing order, customer ask for refund but treated by rude reply," eZone. Accessed 24 June 2022, https://ezone. ulifestyle.com.hk/article/2939498/%E3%80%90%E5%AE%89%E5%BF%83%E5%87% BA%E8%A1%8C%E3%80%91%E9%BB%9E%E9%A4%90%E5%BE%8C%E5%A1% AB%E8%B3%87%E6%96%99%E7%99%BC%E7%8F%BE%E5%80%8B%E8%B3%8 7%E4%BB%BB%E7%9D%87%20%20%E4%BA%8B%E4%B8%BB%E8%A6%81%E 6%B1%82%E9%80%80%E6%AC%BE%E9%81%AD%E5%BE%85%E6%87%89%E6 %83%A1%E5%8A%A3%E5%B0%8D%E5%BE%85.

[12] *Ibid.*

[13] *Ibid.*

[14] *Ibid.*

Nevertheless, unlike using high technology to track COVID-19 cases, using handwritten memos requires extra attention to protect the personal data given by the visitors and prevent data leaking. PCPD concludes three conditions as data protection failure: having visitors share the same memo/notebook, not putting the memo in a storage box and not covering the storage box properly.[15] On 28 October 2021, PCPD received 14 complaints about restaurants from customers for not storing the handwritten memo properly and their personal data were inadvertently disclosed to the public.[16] The debate on using LeaveHomeSafe and handwritten memos is a non-conclusive issue because all visitors have their own right to choose which method to use, until 9 December 2021 when the Hong Kong Government finally announced the compulsory use of LeaveHomeSafe for all scheduled premises in Hong Kong except for peoples under the age of 16 without an accompanying adult and persons aged 65 or above).[17] And, on 27 January 2022, LeaveHomeSafe version 3.0.2 combines the different functions of vaccination record-keeping, which the application will indicate whether the user has been vaccinated or not.

Public Reactions to LeaveHomeSafe

The Hongkong Federation of Youth Groups conducted oral interviews with 806 Hong Kong citizens (aged from 18 to 65), authored case studies

---

[15] *Ibid.*

[16] "Office of the Privacy Commissioner: handwritten memo has high risk than using LeaveHomeSafe," LITENEWS.HK. Accessed 24 June 2024, https://www.litenews.hk/news/13128-%E7%A7%81%E9%9A%B1%E5%85%AC%E7%BD%B2%EF%B9%95%E5%A1%AB%E7%B4%99%E4%BB%94%E7%A7%81%E9%9A%B1%E9%A2%A8%E9%9A%AA%E5%A4%A7%E6%96%BC%E3%80%8C%E5%AE%89%E5%BF%83%E5%87%BA%E8%A1%8C%E3%80%8D.

[17] Schedule premises in Hong Kong including amusement game centre, bathhouse, fitness centre, place of amusement, place of public entertainment, party room, beauty parlour, club-house, club or night club, karaoke establishment, mahjong-tin kau premises, massage establishment, sports premises, swimming pool, hotel or guesthouse, cruise ship, barber shop/hair salon, religious premises, shopping mall, department store, market and supermarket. For details, please visit "Cap. 599F Prevention and Control of Disease (Requirements and Directions) (Business and Premises) Regulation," Hong Kong e-Legislation. Accessed 24 June 2022, https://www.elegislation.gov.hk/hk/cap599F!en-zh-Hant-HK@2022-02-10T00:00:00?INDEX_CS=N&pmc=1&m=1&pm=0.

with 20 Hong Kong youths (aged from 18 to 35) and with 5 scholars between 5 and 27 November 2020.[18] Among these 831 informants, 73.8% of them decided not to use LeaveHomeSafe.[19] They are concerned about the new technology as a means of surveillance of citizens by the Hong Kong Government.[20] The report concluded that the failure of encouraging citizens to use LeaveHomeSafe is that most people do not know how and where their personal data would be used, when the data would be deleted and which parties would be able to obtain the data. The Hongkong Federation of Youth Groups therefore suggested the Hong Kong Government should announce to the public that the use of personal data collected by LeaveHomeSafe on COVID-19 would be restricted to case tracing only and to make sure the data would not/be used for law enforcement, which was a big concern for the those who chose not to use the technology.[21]

In fact, when LeaveHomeSafe was first released on 16 November 2020, the application upset the public for requesting 15 types of rights to access on user's mobile phone, including the two most controversial terms "right to read, edit and delete USB content" and "right to access Wi-Fi connection." The public was worried that LeaveHomeSafe is asking for too much right in accessing items that are related to the privacy of users, and the Office of the Government Chief Information Director Lam Wai-kiu finally made a decision to limit the number of rights to access for LeaveHomeSafe from 15 to 7 (including the cancellation of the two most controversial terms on 28 November 2020) 12 days after the public release of the mobile application.[22]

---

[18] "Impact of universal campaign of fighting COVID-19 on public health," The Hongkong federation of youth groups. Accessed 24 June 2022, https://yrc.hkfyg.org.hk/wp-content/uploads/sites/56/2021/01/YI056_SummaryChi.pdf.

[19] *Ibid.*

[20] *Ibid.*

[21] *Ibid.*

[22] "LeaveHomeSafe requests less access right, Officer of the Government Chief Information Officer hopes more people use the app," Ming Pao. Accessed 24 June 2022, https://news.mingpao.com/pns/%E6%B8%AF%E8%81%9E/article/20201204/s00002/16070 19917102/%E5%AE%89%E5%BF%83%E5%87%BA%E8%A1%8C%E6%B8%9B%E7 %B4%A2%E6%AC%8A%E9%99%90-%E8%B3%87%E7%A7%91%E8%BE%A6%E7 %9B%BC%E6%9B%B4%E5%A4%9A%E4%BA%BA%E7%94%A8.

The HA (Hospital Authority) Employees Alliance (HAEA) did not agree to LeaveHomeSafe as a means to stop the spread of COVID-19 cases in Hong Kong. They set up street counters and distributed promotional flyers on 18 February 2021 to encourage the public to boycott the use of LeaveHomeSafe.[23] From the HAEA perspective, boycotting LeaveHomeSafe is to express the public's distrust of the Hong Kong Government, as the organization believes that the use of LeaveHomeSafe is for surveillance instead of COVID-19 case tracking as the Hong Kong Government claimed.[24] Also, HAEA is not happy with the continuous upgrading of the mobile application with added new "tracking" functions.[25]

LeaveHomeSafe once again was not able to gain the trust of the public who failed to get the source code of the mobile application. A source code for mobile application is a reliable tool for the user to decide rather to use it or not. Knowing the source code allows the public to examine the risk of using the application. In the case of LeaveHomeSafe, the public would like to learn about the source code to check privacy controls on personal data and to make sure the application does not contain Trojan Horse system or spyware. LeaveHomeSafe, owned by the Office of the Government Chief Information Officer (OGCIO), unfortunately, is an application that can never be checked because the government refused to disclose the source code of LeaveHomeSafe. According to OGCIO's director, LeaveHomeSafe was designed by a private company, and it is not

---

[23] "Hospital Authority Employees Alliance asks for boycott on LeaveHomeSafe, Hospital Authority and Officer of the Government Chief Information Officer blame on the false information," On.cc. Accessed 24 June 2022, https://hk.on.cc/hk/bkn/cnt/news/20210219/bkn-20210219115757482-0219_00822_001.html.

[24] *Ibid.*

[25] "Hospital Authority Employees Alliance claims the LeaveHomeSafe mislead the public and Officer of the Government Chief Information Officer is using the app to build up surveillance system," Headline Daily. Accessed 24 June 2022, https://hd.stheadline.com/news/realtime/hk/2003217/%E5%8D%B3%E6%99%82-%E6%B8%AF%E8%81%9E-%E9%86%AB%E7%AE%A1%E5%B1%80%E5%93%A1%E5%B7%A5%E9%99%A3%E7%B7%9A%E6%89%B9%E5%89%B5%E7%A7%91%E5%B1%80%E5%B0%B1-%E5%AE%89%E5%BF%83%E5%87%BA%E8%A1%8C-%E8%AA%A4%E5%B0%8E%E5%B8%82%E6%B0%91%E5%9C%96%E5%BB%BA%E7%AB%8B%E7%9B%A3%E6%8E%A7%E7%B3%BB%E7%B5%B1.

approbated to disclose the code to the public since the government has to protect the intellectual property of the company.[26]

Another disappointment of the public towards the LeaveHomeSafe is focused on its effectiveness in tracing COVID-19 cases. On 27 December 2021, an Omicron outbreak in Hong Kong occurred at a restaurant cluster at Moon Palace restaurant. The "Moon Palace cluster" started from a flight attendant who dined at the Moon Palace restaurant during her stay-home quarantine period. Fifteen people were COVID-19 infected in this cluster and this marked the start of the fifth wave of COVID-19 in Hong Kong. LeaveHomeSafe, however, failed to stop the spread of COVID-19 virus from this "Moon Palace cluster." LeaveHomeSafe recorded the check-in and check-out time and venue of the user, but it does not contain the function of tracking. In the case of the Moon Palace restaurant, there were over 200 customers who visited the restaurant in the same time period in proximity to the COVID-19-infected flight attendant on 27 December 2021.

By 3 January 2022, however, only about 170 customers were con-tactable and about 30 customers were not. To the public's surprise, the Centre of Health Protection obtained the contact of the 170 customers through payment receipts and reservation records provided by Moon Palace restaurant instead of LeaveHomeSafe, the application that was therefore investigated for its capabilities in COVID-19 case tracking. LeaveHomeSafe is not able to stop the spread of the "Moon Palace cluster" apparently, and the public showed disappointment towards the mobile application, including Prof. Yuen Kwok-yung from the Department of Microbiology of the University of Hong Kong. From Prof. Yuen's perspective, the function of LeaveHomeSafe was incomplete and outdated because the application lacks a tracking system, and it was not possible to

---

[26]"Over 2.5 millions download on LeaveHomeSafe, not to open source code as to protect "intellectual property," Ming Pao. Accessed 24 June 2022, https://news.mingpao.com/pns/%E8%A6%81%E8%81%9E/article/20210222/s00001/1613931845636/%E5%AE%89%E5%BF%83%E5%87%BA%E8%A1%8C%E9%8C%84%E9%80%BE250%E8%90%AC%E6%AC%A1%E4%B8%8B%E8%BC%89-%E4%B8%8D%E9%96%8B%E6%94%BE%E5%8E%9F%E5%A7%8B%E7%A2%BC%E7%A8%B1%E3%80%8C%E5%B0%8A%E9%87%8D%E7%9F%A5%E8%AD%98%E7%94%A2%E6%AC%8A%E3%80%8D.

stop the spread of COVID-19 in Hong Kong using only LeaveHomeSafe.[27] Prof Yuen also encouraged the public to allow the government to use phone numbers or smart card data for the purchase (purpose?) of COVID-19 case tracking since LeaveHomeSafe has become unreliable in obtaining contacts and stopping the spread of COVID cases.[28]

In 2022, the public was shocked by the news announced by *FactWire*, a non-profit news agency in Hong Kong, and its discovery of a new fact related to LeaveHomeSafe: the mobile application contains the function of biometric authentication and is able to identify face and location of the user.[29] As the Hong Kong Government does not disclose the source code of the application, it is hard for the public to learn if the biometric authentication function was on while using it. In May 2022, the Hong Kong Government admitted such a function does exist on LeaveHomeSafe ever since it was developed by the developer Cherrypicks Limited, but this function, as claimed by the government, has not been used. To ease the public worries about personal data leak, the government has asked the developer to try to delete biometric authentication function on LeaveHomeSafe.[30]

Leave Home without LeaveHomeSafe

In the *Report on the Survey on Information Technology Usage and Penetration in the Business Sector for 2021* published by the Office of the

---

[27]"LeaveHomeSafe does not contain tracking function, Moon Palace cluster trace is still in process," Yahoo HK News. Accessed 24 June 2022, https://hk.news.yahoo.com/%E6% 9C%AC%E5%9C%B0%E7%BE%A4%E7%B5%84-%E5%AE%89%E5% BF%83%E5%87%BA%E8%A1%8C%E7%84%A1%E8%BF%BD%E8%B9%A4%E5% 8A%9F%E8%83%BD-%E6%9C%9B%E6%9C%88%E6%A8%93%E8%BF%BD%E8% B9%A4%E5%B7%A5%E4%BD%9C%E4%BB%8D%E5%9C%A8%E9%80% B2%E8%A1%8C-051923819.html.
[28]*Ibid.*
[29]"LeaveHomeSafe contains biometric authentication function, Office of the Government Chief Information Officer admitted the fact and plan to delete the function," Yahoo HK News. Accessed 24 June 2022, https://hk.news.yahoo.com/%E3%80%8C%E5%AE%89% E5%BF%83%E5%87%BA%E8%A1%8C%E3%80%8D%E5%85%A7%E7%BD%AE% E4%BA%BA%E8%87%89%E8%AD%98%E5%88%A5%E5%8A%9F%E8%83%BD- %E8%B3%87%E7%A7%91%E8%BE%A6%E6%89%BF%E8%AA%8D-%E7%A8%B 1%E5%BE%9E%E6%9C%AA%E4%BD%BF%E7%94%A8-%E6%AD%A3%E7%A0 %94%E7%A9%B6%E5%88%AA%E9%99%A4-105831448.html.
[30]There is no news regarding the deletion the biometric authentication on LeaveHomeSafe by the time this article is written in June 2022.

Government Chief Information Officer, 68.1% of persons aged 65 and over in Hong Kong own a smartphone.[31] LeaveHomeSafe became a compulsory app for people who lived in Hong Kong on 9 December 2021 when the government announced the discontinuation of handwritten memo system in replacing using LeaveHomeSafe. From now on, all people need to use LeaveHomeSafe to enter any of the scheduled premises. So, how did the rest of the 31.9% of persons aged 65 and over live without LeaveHomeSafe?

MingPao Weekly conducted an interview with a 72-year-old hawker Fok and ask about her life without a smartphone.[32] Fok feels forgotten by the society in the past two years ever since COVID-19 outbreak in Hong Kong. LeaveHomeSafe is one of the biggest obstacles to living in Hong Kong because she is one of the 31.9% who does not have extra expenses of owning a smartphone. Although the government allows people who aged over 65 years old to use handwritten memos to replace LeaveHomeSafe, Fok avoids visiting scheduled premises as she does not want to spend time lining up for memo filling just to enter the market. Now Fok is not able to enter the markets, supermarkets or shopping malls, and she has to spend extra time to walk to the nearby stores that do not require LeaveHomeSafe. Fok's case reflects the reality faced by the elderly in Hong Kong nowadays as poverty and lacking technological knowledge prevented some people from living a normal life.

Am730 published a feature on the day when LeaveHomeSafe became compulsory for wet market visit on 1 November 2021.[33] According to the

---

[31] "Thematic Household Survey Report No. 73," Office of the Government Chief Information Officer. Accessed 24 June 2022, https://www.ogcio.gov.hk/tc/about_us/facts/doc/householdreport2021_73.pdf.

[32] "Not able to use the Internet, no money to buy mobile phone, 72-year-old female faces difficulties in daily life, handwritten memo is troublesome, difficult to apply for consumption voucher: just want to stay healthy and not to deal with too much issues," Ming Pao Weekly. Accessed 24 June 2022, https://www.mpweekly.com/culture/%E9%95%B7%E8%80%85-%E6%95%B8%E7%A2%BC%E8%83%BD%E5%8A%9B-%E8%80%81%E4%BA%BA%E5%AE%B6-194804.

[33] "Seniors allow to enter the wet market without LeaveHomeSafe or handwritten memo, long queue outside the wet market and complained by the citizens," am730. Accessed 24 June 2022, https://www.am730.com.hk/%E6%9C%AC%E5%9C%B0/%E5%AE%89%E5%BF%83%E5%87%BA%E8%A1%8C-%E6%9C%89%E9%95%B7%E8%80%85%E7%84%A1%E6%8E%83%E7%A2%BC%E7%84%A1%E5%A1%AB%E7%B4%99%E4%BB%94%E4%BB%8D%E5%87%86%E5%85%A5%E8%A1%97%E5%B8%82-%E9%96%80%E5%A4%96%E6%8E%92%E9%95%B7%E9%BE%8D%E8%A1%97%E5%9D%8A%E6%89%B9%E6%93%BE%E6%B0%91/292611.

news report, some seniors found out LeaveHomeSafe was needed for all wet markets only when they arrived at the entrance. There were queues outside the wet markets and many of them were lining up at the counter to fill in personal data on the handwritten memo. However, the report shows that not many of the staff check the eligibility of those who chose handwritten memos over LeaveHomeSafe.[34] Yeung, who visited the wet market with handwritten memos, expressed his disappointment with the arrangement.[35] He felt upset with the unpleasant visit to the wet market and thinks that either LeaveHomeSafe or leaving personal data by hand-written memos are time-wasting measures that are not able help stop COVID-19 spread in the society but only bring more trouble for its citizens.[36]

The Social Innovation and Entrepreneurship Development Fund (SIE) was set up by the government for supporting poverty relief in Hong Kong. SmartConnect is a project by SIE that tries to ease the worries of the elderly by using advanced digital technologies. SmartConnect project organizes "Digital Mobile Classroom" to reach out to the elderly "living in remote areas and provide them with knowledge of digital technologies and technical support."[37] Teaching the seniors how to use LeaveHomeSafe is on the agenda of SIE. SIE creates YouTube videos to teach seniors how to use LeaveHomeSafe and the video was viewed over 7,000 times.[38]

Knowing the difficulties faced by the elderly in Hong Kong, private sector started to provide help to the needy, and Hong Kong Jockey Club is one of them. The Hong Kong Jockey Club Charities Trust donated HK$27 million in December 2021 to launch the Jockey Club Digital Support Project for the Elderly project in collaboration with 12 NGO partners and four mobile network service providers.[39] The project aims to benefit the elderly in the community who cannot afford a smartphone and

---

[34] *Ibid.*

[35] *Ibid.*

[36] *Ibid.*

[37] "Smart Connect," Social Innovation and Entrepreneurship. Accessed 24 June 2022, https://www.sie.gov.hk/sie/en/our-work/funded-ventures/SmartConnect.page?

[38] "Know more about "LeaveHomeSafe" — tutorial #01 for seniors," SmartConnect. Accessed 24 June 2022, https://www.youtube.com/watch?v=xYsK262vr_M.

[39] "Jockey Club Digital Support Project for the Elderly," Jockey Club. Accessed 24 June 2022, https://jc-elderly.digital/home/.

are disconnected from the digital world. The Hong Kong Jockey Club used the donations to purchase smartphones and installed LeaveHomeSafe for the elderly before giving the phones to them. Also, this phone-giving project includes a free 12-month basic mobile data service so that the user can use it for free. The Hong Kong Jockey Club partners-operated elderly community centres and neighbourhood elderly centres will help to identify eligible the marginalized elderly, and these approximately 20,000 identities elderlies will be taught by the volunteers on how to use a smartphone, connect to mobile data and, most importantly, use LeaveHomeSafe.[40]

Is It Safe to Leave Home?

LeaveHomeSafe serves the purpose of tracing the records of people who have been to the same place with COVID-19-infected individuals. But, why did the Hong Kong citizens not feel safe using the mobile application? Apparently, the hidden source code bears the responsibility for the LeaveHomeSafe failure. The government should carry out a more thoughtful check on the source code before announcing the app to the public, as even the vice director of OGCIO admitted that the Department was not aware of the function of biometric authentication on LeaveHomeSafe.[41] Most importantly, either the developer Cherrypicks Limited or the Hong Kong Government failed to explain the reason for installing biometric authentication function on LeaveHomeSafe to the public, which only led to the failure of the mobile application in gaining trust from the public. In fact, it is not an easy task for the government to encourage the public to use LeaveHomeSafe as even the authorities did not know about the content of the mobile application and failed to show quality control by OGCIO. The case of LeaveHomeSafe reflects the ability of the Hong Kong Government in promoting the use of high technology in the society. It is, indeed, a good lesson for the government to understand how the public view using high technology in daily life.

---

[40] *Ibid.*

[41] "LeaveHomeSafe faces trouble, how to development innovation and technology centre," HK01. Accessed 24 June 2022, https://www.hk01.com/01%E8%A7%80%E9%BB%9E/7 66904/%E5%AE%89%E5%BF%83%E5%87%BA%E8%A1%8C%E5%B1%A2%E5%B 1%A2%E7%A2%B0%E5%A3%81-%E8%AB%87%E4%BD%95%E7%99%BC%E5% B1%95%E5%89%B5%E7%A7%91%E4%B8%AD%E5%BF%83.

If turning Hong Kong into a smart city is still the government's agenda, there is a long way to go in order to get the entire society and its people prepared for this big project. This chapter completes the section on Hong Kong analyses. The following chapter is the conclusion of this volume, followed by the epilogue.

Bibliography

"Cap. 599F Prevention and Control of Disease (Requirements and Directions) (Business and Premises) Regulation," Hong Kong e-Legislation. Accessed 24 June 2022, https://www.elegislation.gov.hk/hk/cap599F!en-zh-Hant-HK@2022-02-10T00:00:00?INDEX_CS=N&pmc=1&m=1&pm=0.

"Customers failed to use LeaveHomeSafe and filled in wrong personal data, Mongkok Hot pot restaurant owner is fined HK$8,000," Bastille Post. Accessed 24 June 2022, https://www.bastillepost.com/hongkong/article/9639102-%E9%A3%9F%E5%AE%A2%E7%84%A1%E7%94%A8%E5%AE%89%E5%BF%83%E5%87%BA%E8%A1%8C%E5%8F%8A%E4%BA%82%E5%A1%AB%E8%B3%87%E6%96%99-%E6%97%BA%E8%A7%92%E7%81%AB%E9%8D%8B%E5%BA%97%E8%B2%A0%E8%B2%AC%E4%BA%BA%E5%88%A4.

"Handwritten-memo-visitors use LeaveHomeSafe for the first time, feel safer to install on the extra mobile phone," CitizenNews. Accessed 24 June 2022, https://www.youtube.com/watch?v=gqF0m3LWPQY.

"Hospital Authority Employees Alliance asks for boycott on LeaveHomeSafe, Hospital Authority and Officer of the Government Chief Information Officer blame on the false information," On.cc. Accessed 24 June 2022, https://hk.on.cc/hk/bkn/cnt/news/20210219/bkn-20210219115757482-0219_00822_001.html.

"Impact of universal campaign of fighting COVID-19 on public health," The Hongkong federation of youth groups. Accessed 24 June 2022, https://yrc.hkfyg.org.hk/wp-content/uploads/sites/56/2021/01/YI056_SummaryChi.pdf.

"Jockey Club Digital Support Project for the Elderly," Jockey Club. Accessed 24 June 2022, https://jc-elderly.digital/home/.

"Know more about "LeaveHomeSafe" – tutorial #01 for seniors," SmartConnect. Accessed 24 June 2022, https://www.youtube.com/watch?v=xYsK262vr_M.

"Launch of "LeaveHomeSafe" COVID-19 exposure notification mobile app," The Government of the Hong Kong. Accessed 24 June 2022, https://www.info.gov.hk/gia/general/202011/11/P2020111100367.htm?fontSize=1.

"LCQ8: "LeaveHomeSafe" mobile application," The Government of the Hong Kong. Accessed 24 June 2022, https://www.info.gov.hk/gia/general/202012/16/P2020121600222.htm?fontSize=1.

"LeaveHomeSafe does not contain tracking function, Moon Palace cluster trace is still in process," Yahoo HK News. Accessed 24 June 2022, https://hk.news.yahoo.com/%E6%9C%AC%E5%9C%B0%E7%BE%A4%E7%B5%84-%E5%AE%89%E5%BF%83%E5%87%BA%E8%A1%8C%E7%84%A1%E8%BF%BD%E8%B9%A4%E5%8A%9F%E8%83%BD-%E6%9C%9B%E6%9C%88%E6%A8%93%E8%BF%BD%E8%B9%A4%E5%B7%A5%E4%BD%9C%E4%BB%8D%E5%9C%A8%E9%80%B2%E8%A1%8C-051923819.html.

"LeaveHomeSafe faces trouble, how to development innovation and technology centre," HK01. Accessed 24 June 2022, https://www.hk01.com/01%E8%A7%80%E9%BB%9E/766904/%E5%AE%89%E5%BF%83%E5%87%BA%E8%A1%8C%E5%B1%A2%E5%B1%A2%E7%A2%B0%E5%A3%81-%E8%AB%87%E4%BD%95%E7%99%BC%E5%B1%95%E5%89%B5%E7%A7%91%E4%B8%AD%E5%BF%83.

"LeaveHomeSafe requests less access right, Officer of the Government Chief Information Officer hopes more people use the app," Ming Pao. Accessed 24 June 2022, https://news.mingpao.com/pns/%E6%B8%AF%E8%81%9E/article/20201204/s00002/1607019917102/%E5%AE%89%E5%BF%83%E5%87%BA%E8%A1%8C%E6%B8%9B%E7%B4%A2%E6%AC%8A%E9%99%90-%E8%B3%87%E7%A7%91%E8%BE%A6%E7%9B%BC%E6%9B%B4%E5%A4%9A%E4%BA%BA%E7%94%A8.

"Not able to use the Internet, no money to buy mobile phone, 72-year-old female faces difficulties in daily life, handwritten memo is troublesome, difficult to apply for consumption voucher: just want to stay healthy and not to deal with too much issues," Ming Pao Weekly. Accessed 24 June 2022, https://www.mpweekly.com/culture/%E9%95%B7%E8%80%85-%E6%95%B8%E7%A2%BC%E8%83%BD%E5%8A%9B-%E8%80%81%E4%BA%BA%E5%AE%B6-194804.

"Office of the Privacy Commissioner for Personal Data received 49 complain cases on handwritten memo, 15 cases are found, LeaveHomeSafe is better," CABLE TV & CABLE News. Accessed 24 June 2022, https://www.youtube.com/watch?v=bqMRv9oAIq8.

"Over 2.5millions download on LeaveHomeSafe, not to open source code as to protect "intellectual property," Ming Pao. Accessed 24 June 2022, https://news.mingpao.com/pns/%E8%A6%81%E8%81%9E/article/20210222/s00001/1613931845636/%E5%AE%89%E5%BF%83%E5%87%BA%E8%A1%8C%E9%8C%84%E9%80%BE250%E8%90%AC%E6%AC%A1%E4%B8%8B%E8%BC%89-%E4%B8%8D%E9%96%8B%E6%94%BE%E5%8E%9F%E5%A7%8B%E7%A2%BC%E7%A8%B1%E3%80%8C%E5%B0%8A%E9%87%8D%E7%9F%A5%E8%AD%98%E7%94%A2%E6%AC%8A%E3%80%8D.

"Seniors allow to enter the wet market without LeaveHomeSafe or handwritten memo, long queue outside the wet market and complained by the citizens," am730. Accessed 24 June 2022, https://www.am730.com.hk/%E6%9C%A C%E5%9C%B0/%E5%AE%89%E5%BF%83%E5%87%BA%E8%A1 %8C-%E6%9C%89%E9%95%B7%E8%80%85%E7%84%A1%E6%8E%8 3%E7%A2%BC%E7%84%A1%E5%A1%AB%E7%B4%99%E4%BB%94 %E4%BB%8D%E5%87%86%E5%85%A5%E8%A1%97%E5%B8%82- %E9%96%80%E5%A4%96%E6%8E%92%E9%95%B7%E9%BE%8D%E 8%A1%97%E5%9D%8A%E6%89%B9%E6%93%BE%E 6%B0%91/292611.

"Smart Connect," Social Innovation and Entrepreneurship. Accessed 24 June 2022, https://www.sie.gov.hk/sie/en/our-work/funded-ventures/Smart Connect.page?

"Thematic Household Survey Report No. 73," Office of the Government Chief Information Officer. Accessed 24 June 2022, https://www.ogcio.gov.hk/tc/ about_us/facts/doc/householdreport2021_73.pdf.

Part V

Conclusion and Epilogue

China's Space Programme: The Final Frontier

Conclusion: This volume started with China's tech development from a world history with the case study of maritime navigational technologies which can provide a longitudinal historical development of one particular Chinese tech sector. From this case study, it is possible to detect a long history of technological development in China. To understand how contemporary Chinese companies transformed the country into a tech superpower, the subsequent chapter examined how contemporary China develop its technologies. After its survey on contemporary Chinese technological development, this volume then proceeded to examine selected specific case studies in the following chapters, including how China powered its technological development in heavy industries and other sectors through the use of coal energy.

Having discussed mostly manufacturing industries up to that point, the subsequent case study focused on a non-manufacturing sector to examine how Chinese companies applied technologies in the agricultural sector. Having detailed China's agricultural modernization and technological use, it was followed by a comparative perspective examining the agricultural sub-sector of eco-feed by comparing its development with those of neighbouring locations of Japan and Taiwan. In this chapter, China's development in high-value-added eco-feed for rearing chickens was compared with Japan's case study. It is a comparative study of how far China has developed its technologies in this sector. Some references are also made to "Greater China" developments like Taiwan, whose consumers like Japan and increasingly China are demanding higher-quality premium organic foods for their consumption.

After examining China's manufacturing and agricultural firms/companies' use of technologies domestically, the final chapter in the section on China focused on the external application of its technologies for other ASEAN/East Asian economies. With rapid technological development, China and its state and private sectors are now ready to assist other countries with their own developments. Perhaps the region that stands to gain from this is the nearest region to China, the neighbouring region of Southeast Asia. This chapter examined such roles in the context of ASEAN's desire for connectivity.

The second major section in this volume focused on the Hong Kong case study. The first chapter in this section examined how, within China, the pioneer in the capitalist experiment and former tiger economy HKSAR is attracting tech companies with robust R&D strengths, infrastructure support for the tech industry and track record for IPR protection.[1] Having discussed domestic developments in the Hong Kong tech sector, this volume then proceeded to examine the wider macro context of Hong Kong companies/firms' role in the context of the GBA regionalism. This chapter examined Hong Kong's GBA macro-region for its techno-economic future. The nexus between Hong Kong and China's development were detailed here.

Having discussed Hong Kong tech enterprises' role in developing their technological sector and the embeddedness of Hong Kong developments in the context of GBA regionalism, the following chapter focused on specific sectoral case studies, including maritime technologies. The chapter on the case study of "Hong Kong's Maritime Technologies" detailed the development of Hong Kong as a global port city with sophisticated container traffic and shipping activities, including the roles played by its shipping companies and public policy.

The section on Hong Kong wraps up with the chapter on "Utilization of Advanced Technology during the COVID-19 pandemic in Hong Kong" which details the technologies that Hong Kong used to mitigate COVID-19

---

[1]Hawksford, "Why Is Hong Kong a Unique Place for Tech Companies?" dated 7 April 2021 in GuideMeHongKong Hawksford (downloaded on 7 April 2021). Available at https://www.guidemehongkong.com/in-the-news/2021---why-is-hong-kong-a-unique-place-for-tech-companies.

coronavirus pandemic. This chapter updated Hong Kong's technological developments up to speed with contemporary information. Finally, this conclusion and the epilogue for this volume examined China and the scientific final frontiers in space exploration. The final chapter in this volume looks at the development of science and technology in the arena of space exploration contextualized in the progress that the Chinese space programme has made over the years.[2]

The origins of China's space programme dates back to 1956 in the research facilities of the Fifth Academy of the Ministry of National Defence (now known as China Aerospace Science and Technology Corporation).[3] Its initial goals were ideologically spurred by Mao Zedong's 1958 speech that China should catch up with the Soviet Union and the United States in launching a satellite into space, a feat accomplished on 24 April 1970 with the successful orbital launch of Dong Fang Hong 1.[4]

A decade thereafter in the 1980s, much of China's space efforts were spent building their own launch sites in Xichang (Sichuan province) and Taiyuan (Northern China) before developing the Shenzhou space module in the 1990s and sending China's first taikonaut[5] Yang Liwei into space in 2003. Much of this development was indigenous with Soviet help in 1960 to develop the CZ (or Changzheng/Long March) rocket and the 1996 procurement of Russian technologies.[6]

Officially, China encourages peaceful utilization of outer space, a goal announced by China National Space Administration spokesperson Li Guoping in 2018. It has signed 121 cooperation agreements with 37 countries and four international organizations and worked with countries like Brazil (earth resources satellites), Algeria (communication

---

[2]The author first broached this topic in: Lim, Tai Wei, "Japan's Space Programme; The Final Frontier" dated 1 October 2020 in NUS EAI Background Brief (Singapore: NUS EAI), 2020. This section of the epilogue contains materials from this source.

[3]Zhu, Sirui, "China's Long March to Space" dated 27 December 2019 in Reuters (downloaded on 27 December 2019). Available at https://graphics.reuters.com/SPACE-EXPLORATION-MOON/0100B0BH0NZ/index.html.

[4]*Ibid.*

[5]Chinese astronauts are known as 'taikonauts'.

[6]New Scientist and AFP, "Timeline: China's spaceflight history" dated 12 October 2005 in New Scientist (downloaded on 1 January 2020). Available at https://www.newscientist.com/article/dn8144-timeline-chinas-spaceflight-history/.

satellite launch in 2017), Pakistan/Iran/Turkey/Peru (satellites that track floods and forest fires), Thailand/Laos/Burma/Cambodia (remote-sensing information platform), among others.[7]

China Academy of Launch Vehicle Technology, a major manufacturer of carrier rockets, is also working on turning advanced space technologies into commercialized products, e.g. industrial-grade space-tech robotic arms for holding things with a precision of 0.1 mm.[8] China's domestic industrial robot market is large enough to absorb these technologies derived from space equipment; the industrial parks in Chongqing and Beijing, for instance, have become ready customers.[9]

China is also working with other Asia-Pacific countries in managing outer space tracking and surveillance data. China's membership in the Asia-Pacific Space Cooperation Organisation and its surveillance initiative known as the Asia-Pacific Ground-Based Optical Space Object Observation System has seen China generously supplying telescopes to Peru, Pakistan and Iran to track objects in low Earth orbit and geostationary orbit; such data are then fed back to the Chinese Academy of Science's National Astronomical Observatory.[10]

Indonesia is working with China. China has built the Palapa-N1 communication satellite for Indonesia at the Xichang Satellite Launch Centre in Sichuan in a deal inked between China Great Wall Industry Corp (under the umbrella of China Aerospace Science and Technology Corp) and Indonesian Palapa Satelit Nusantara Sejahtera in May 2017.[11] The Indonesian Palapa-N1 satellite was constructed on a Dongfanghong-4 platform built by the China Academy of Space Technology with a Long

---

[7]*Xinhua News Agency*, "China strengthens international space cooperation" dated 23 April 2018 in Space Daily (downloaded on 23 April 2018). Available at https://www.spacedaily.com/reports/China_strengthens_international_space_cooperation_999.html.

[8]*Xinhuanet*, "China Focus: China's space tech-based robots find wider application on Earth" dated 22 August 2019 in Xinhuanet (downloaded on 22 August 2019). Available at http://www.xinhuanet.com/english/2019-08/22/c_138329594.htm.

[9]*Ibid.*

[10]Stokes, Mark, Gabriel Alvarado, Emily Weinstein and Ian Easton, China's Space and Counterspace Capabilities and Activities (The US–China Economic and Security Review Commission), 2020, pp. 27–28.

[11]*Xinhua*, "China to launch communication satellite for Indonesia" dated 1 April 2020 in Xinhuanet (downloaded on 1 April 2020). Available at http://www.xinhuanet.com/english/2020-04/01/c_138937376.htm.

March-3B carrier rocket mounted by the China Academy of Launch Vehicle Technology that delivers the satellite into space.[12]

China is also working with other Asian countries and beyond the region. On 11 January 2018, China Great Wall Industry Corporation worked with the Royal Cambodia Group to build the Techno 1 communications satellite for use by the Cambodian Broadcasting Service Company, Cellcard (Cambodia's third biggest mobile operator) and the Cambodian government to manage natural disaster, national security issues and other government services.[13] China also has a longstanding space programme with Iran. On 7 September 2008, an Iranian-Thai-Chinese collaborative scientific research satellite, Environment-1, was launched into space on the back of a Chinese rocket.[14]

Another longstanding partner with China is Venezuela. In October 2008, China launched Venezuela's first Venesat-1 communications satellite on a Chinese Long March 3B from the Xichang Satellite Launch Centre and in September 2012, another Chinese Long March 2D rocket launched Venezuela's first remote-sensing satellite, the VRSS-1, into space from the Jiuquan.[15] Venezuela's second VRSS-2 remote-sensing satellite was launched into space on a Long March 2D rocket from Jiuquan Satellite Launch Centre; it was designed by the China Academy of Space Technology in Beijing with a panchromatic/multi-spectral high-resolution imager and an infrared camera for use in land resources inspection, environmental protection, disaster monitoring/relief, crop yield estimation and urban planning.[16]

China was also Pyongyang's alternative source of space-based technological hardware. According to Reuters, North Korea had, in the recent

---

[12] *Ibid.*

[13] Goh, Deyana, "China to build and launch Cambodia's first satellite" dated 12 January 2018 in SpaceTech Asia (downloaded on 12 January 2018). Available at https://www.spacetechasia.com/china-to-build-and-launch-cambodias-first-satellite/.

[14] Hanna, Andrew, "Iran's Ambitious Space Program" dated 29 July 2020 in United States Institute of Peace (USIP) website (downloaded on 29 July 2020). Available at https://iranprimer.usip.org/index.php/blog/2020/jun/23/iran%E2%80%99s-ambitious-space-program.

[15] Zhao, Lei, "China launches second remote-sensing satellite for Venezuela" dated 10 October 2017 in China Daily (downloaded on 10 October 2017). Available at https://www.chinadaily.com.cn/china/2017-10/10/content_33058430.htm.

[16] *Ibid.*

past, access to Chinese space technologies. The DPRK's transporter-erector-launcher vehicle that transports ballistic missiles and then fires them vertically was cited in the UN's 2013 report as the exact model as that of China's Hubei Sanjiang Space Wanshan Special Vehicle Company, a subsidiary of China Aerospace Science and Industry Corp, a state-owned enterprise (SOE) that manufactures the Shenzhou space rockets.[17] The mobile nature of the missile-launching vehicle makes detection difficult for the United States and its allies.

Not all of China's space capabilities services or delivers for non-Western countries or America's foes. China has developed joint programmes with some European Union countries. China and France have co-developed an ocean-observing satellite to study climate change while Italy and China have built a seismic-electromagnetic satellite to study seismic precursors to track and forecast earthquakes.[18]

China's Beidou system is a remarkable Chinese space achievement. Beidou will enable China to be self-dependent for earth observation, navigation and remote sensing capabilities and it can generate returns from consumers using its services to the tune of US$298 billion dollars (RMB2 trillion) with its seven high-resolution satellites of this type in orbit.[19] It can be a commercial windfall for the Chinese space industry.

However, the US government analyses such capabilities differently, given their dual-use nature. Their extensive report prepared for the US–China Economic and Security Review Commission noted that the PLA can hit targets at greater ranges due to PNT services provided by Beidou and greatly widens the operational range of Chinese strategic forces with satellite communications support for long-distance nuclear submarine navigation, strategic bombers and global missile deployments.[20]

When China's anti-satellite test (ASAT) disintegrated its old satellite in space in January 2007, it sparked the United States and its alliance partners to reconsider ways to protect their assets in space. The White

---

[17]Pearson, James and Jack Kim, "North Korea appeared to use China truck in its first claimed ICBM test' dated 4 July 2017 in Reuters (downloaded on 4 July 2017). Available at https://www.reuters.com/article/us-northkorea-missiles-china-truck-idUSKBN19P1J3.

[18]*Xinhua News Agency*, *op. cit.*

[19]Stokes, Mark, Gabriel Alvarado, Emily Weinstein, and Ian Easton, China's Space and Counterspace Capabilities and Activities, p. 11.

[20]*Ibid.*, p. 20.

House statement given by the NSC reads the following: China's "development and testing of such weapons is inconsistent with the spirit of cooperation that both countries aspire to in the civil space area."[21] The US alliance (Five Eyes, the EU, South Korea/Japan) is concerned that China's Beidou, for example, could pose a security risk by allowing the owner to track users by inserting malware in navigation signals or messaging function in satellite communication and then use the system in non-disclosed ways.[22] The United States is far more worried about the Russian than the Chinese space programmes at the moment. The United States is likely to tap into its allies' expertise to develop satellites that can monitor its rivals' space programmes and weapon systems.

The Global Positioning Navigation System[23]: Given the dual-use importance of satellite technologies, it may be useful to examine China's global positioning tech capabilities here.

Global History: The US set up the inaugural Global Positioning System (GPS) in 1978 and attained international usage in 1995, Russians followed up with their Global Navigation Satellite System (GLONASS) in 1982 and then globalized in 1996, European Space Agency's Galileo started in 2005 and reached semi-operational status in 2019 (without launching all 30 satellites) while India/Japan only have regional coverage.[24] Regardless of the system, they are all based on beaming radio signals to devices on the ground with signals from minimally four satellites to calibrate the receiver's location, direction and velocity.[25] GPS (American), GLONASS (Russian) and Galileo (European) function as

---

[21] Broad, William J. and David E Sanger, "China Tests Anti-Satellite Weapon, Unnerving U.S." dated 18 January 2007 in *New York Times* (downloaded on 1 January 2020). Available at https://www.nytimes.com/2007/01/18/world/asia/18cnd-china.html.

[22] Halappanavar, Abhilash, "The final satellite in the BeiDou system completes an undertaking 20 years in the making" dated 26 June 2020 in *The Diplomat* (downloaded on 26 June 2020). Available at https://thediplomat.com/2020/06/chinas-answer-to-gps-is-now-fully-complete/.

[23] Some materials in this chapter are derived from the author's manuscript on navigational history from a maritime silk road publication.

[24] Andrews, Mark, "Navigating the World" dated 15 March 2021 in the Cheung Kong Graduate School of Business (CKGSB) Knowledge (downloaded on 15 March 2021). Available at https://english.ckgsb.edu.cn/knowledges/china-global-navigation-system-beidou/.

[25] *Ibid.*

beacons that shoot out signals that can be received by billions of devices for triangulating their precise locations on Earth.[26]

Originally a military application, GPS was only available for civilian usage after USSR fighter jets shot down a Korean Airlines Flight 007 in 1983 and killed its 269 passengers on board when the plane flew into Soviet airspace because of navigation guidance error.[27] The US then found out that the economic benefits of the GPS were stronger than the military returns, and free PNT services can facilitate more technological applications in a globalized world like autonomous vehicle systems.[28] Today, GPS is integrated with the modern economy, daily lifestyles and military applications. Interestingly, after two decades of benchmarking and challenging the US GPS system, the Chinese system appear to be more extensive and larger than that of the US. According to Chinese research firm Qianxun SI's report, BeiDou's satellites were observed more frequently than GPS satellites in most parts of the world while Chinese state media Xinhua News Agency reported on 3 July 2020 that Beidou has 500 million subscribers for its precision positioning services.[29]

The Chinese System: Decoupling: China started constructing its own global navigational satellite Beidou (translated literally into "Big Dipper" in English, a constellation often used for navigational guidance) system on 23 June 2020 in outer space through satellite launches and it was perhaps the most ambitious alternative system to the American Global Positioning System (GPS).[30] There is a strategic angle to the system as China did not want to compromise its national security by tapping into the US GPS system operated by the US armed forces for its supply positioning, navigation and timing (PNT) information.[31]

In the 1996 Taiwan Strait Crisis when China was furious at what it perceived as moves towards independence in Taiwan, China fired three

---

[26]Xie, John, "China's Rival to GPS Navigation Carries Big Risks" dated 8 July 2020 in Voice of American (VOA) (downloaded on 8 July 2020). Available at https://www. voanews.com/a/east-asia-pacific_voa-news-china_chinas-rival-gps-navigation-carries-big-risks/6192460.html.

[27]Andrews, Mark, *op. cit.*

[28]*Ibid.*

[29]Xie, John, *op. cit.*

[30]Andrews, Mark, *op. cit.*

[31]*Ibid.*

missiles to coordinate on the Taiwan Strait as a stern warning and only one of the missiles (the first one) hit its target.[32] The Chinese Global Positioning System was therefore conceptualized in the 1990s when the Chinese alleged that, when they fired three missiles into the ocean near Taiwan in 1996 and lost communication with two of them, the US cut off the GPS signals for the missile guidance.[33]

The PLA was anxious that it was exposed to jamming or spoofing by the Americans in battle situations if they continued to rely on the US GPS system, making their weapons and communications systems unusable.[34] Commissioner on the US–China Economic and Security Review Commission Larry Wortzel articulated the following:

> Keep in mind that signals can be jammed, cutting off access to PNT, or spoofed, sending a false location or direction.... Having the capability to receive a variety of PNT signals and process them provides alternatives in the event one country's system is jammed, spoofed or taken offline for any reason.[35]

Blaine Curcio, a space and satcom industry consultant, argued that Beidou enhances pro-decoupling arguments because it can contribute to the deteriorating US–China relations, a point that Namrata Goswami, an independent analyst and author specializing in space policy, concurred the following:

> China's aim to establish an independent navigation system is about decoupling from dependence on foreign navigation systems.... The need for it arises even further with protectionist measures and the penchant to cut communication signals in times of conflict [and that Chinese companies targeted by US sanctions can now easily switch to BeiDou without fearing losing business].[36]

Deteriorating relations with the US spurred China to be more self-reliant tech-wise with Beidou as part of "The Two Pillars of a Great

---

[32] Xie, John, *op. cit.*
[33] Andrews, Mark, *op. cit.*
[34] *Ibid.*
[35] *Ibid.*
[36] *Ibid.*

Power" alongside 5G telecommunications and one way to reach is self-reliance is to ramp up R&D budgets, with China now making up 23% of international R&D expenditure compared with 25% for the US in 2018.[37] Namrata Goswami, an independent analyst and author specializing in space policy, opined the following:

> Learning a lesson by getting key technologies cut off due to US national security concerns, China will make self-reliance in core technologies a priority, and as in other sectors including 5G, space navigation and telecoms, will work toward self-reliance.... This will become the top policy focus given China is the world's largest consumer of computer chips. Such efforts are visible with increased funding in semiconductor research, increasing job opportunities and high salaries drawing outside talent, all with an aim to develop 70% of chips within the country by 2025.[38]

After about 25 years since 1996 when China accused the US of turning off the GPS signals for guiding its warning missiles against perceived Taiwanese separatists, Beidou has overtaken the US GPS in size and scale, and, by the end of June 2020, there were 35 operational Beidou satellites, as opposed to 31 for the US GPS system.[39] If the rivalry persists, in the most extreme case, the world may break into two GPS spheres. Heath Sloane, a scholar at the Yenching Academy of Peking University, projected the following:

> Widespread integration of BeiDou across the Belt and Road [a global development strategy adopted by the Chinese government in 2013] will ostensibly end a member nation's reliance on the American military-run GPS network…Torn between rival networks, the world may soon be bifurcated into GPS or BeiDou camps.[40]

Beidou and Its Progression: Beidou started off as one of 16 science and technology (S&T) major initiatives under China's 2006–2020 medium- to long-term planning for tech development with an aim to overcome US

---

[37] *Ibid.*
[38] *Ibid.*
[39] Xie, John, *op. cit.*
[40] *Ibid.*

dominance in satellite navigation and accelerate China's rise in global power, as Namrata Goswami, an independent analyst and author specializing in space policy, explained the following:

> An independent BeiDou offers a strong hold on global infrastructure and rulemaking.... This enables China to challenge the centrality of the United States to form partnerships and alliances and to control the standards for 5G, information technology, mobile devices, self-driving cars and drones, and the broader Internet of Things (IoT).[41]

Beidou is consistently conceptualized/hailed as "the biggest" Chinese aerospace programme, and in the last 2.5 years of its plan to launch 35 satellites between 2018 and 2020, 300,000 scientists/engineers from drawn 400 research institutions/firms were mobilized.[42] Namrata Goswami, an independent analyst and author specializing in space policy, noted Beijing sunk in approximately US$9 billion to build the system and Beidou is currently the biggest of the four international navigation systems:

> In 1996, China decided to build its own navigation system, to be completed within 25 years, to establish truly independent military command and control, and precision missile guidance and tracking.[43] [Other sources like Voice of America (VOA) noted that Beidou was officially initiated in 1994.[44]]

In its third phase, Beidou Phase 1 is made up of four satellites fired in 2000–2003 with coverage only in China and some Asian regions, Phase 2 included 14 satellites released into space from 2004 to 2012 now able to reach the entire Asia Pacific, Phase 3 (BDS3) put in place a satellite constellation with global reach.[45] After over two decades, China finished the construction of its satellite navigation system on 30 June 2020 when Beidou's 35th satellite attained geostationary orbit successfully and the last Beidou-3 satellite was fired into space on a Long March-3B rocket

---

[41] Andrews, Mark, *op. cit.*
[42] Xie, John, *op. cit.*
[43] Andrews, Mark, *op. cit.*
[44] Xie, John, *op. cit.*
[45] Andrews, Mark, *op. cit.*

from Xichang Satellite Launch Center in Sichuan China on 23 June 2020.[46] Heath Sloane, Yenching Scholar at Peking University, indicated the following: "BeiDou comprises the largest constellation, totalling 35 operational satellites.... This outnumbers other systems by at least four satellites. Notwithstanding the truism that positional accuracy depends on the number of satellites, it isn't the only metric that counts."[47]

A 2014 *China Daily* article mentioned that Beijing intends to construct a network of ground stations in other parts of Asia, including 1000 in Southeast Asia.[48] After attaining live transmission less than half a year later, *China Daily* reported that the system was usable in above 120 countries/regions while China Satellite Navigation Office reports that approximately 70% of hand-phones in China, 6.2 million taxis, buses and lorries are utilizing Beidou.[49] By July 2020, China's state-owned mass media announced Beidou the system was utilized by more than 50% of the globe with its navigation products exported to more than 120 countries.[50]

According to Namrata Goswami, an independent analyst and author specializing in space policy, open-source data indicated Beidou's signal accuracy appears to be better than GPS (0.41 metre as opposed to a 0.5 metre average for GPS), augmented by 155 framework reference stations and more than 2200 regional stations in China, which makes accuracy precise to the centimetre (cm) calibration.[51] BeiDou is a two-way communication system that identifies receivers' locations while BeiDou-compatible equipment sends data back to the satellites (including in text of up to 1200 Chinese characters), as Chinese news channel *CCTV* revealed in June 2020: "In layman's terms, you can not only know where you are through BeiDou but also tell others where you are through the system."[52]

Blaine Curcio, a space/satcom industry consultant, added the following: "But BeiDou has certain features that GPS does not, such as a short messaging system" and a two-way function where a receiving device can transmit a 1200-character message (only the much-smaller French SPOT

---

[46] Xie, John, *op. cit.*

[47] Andrews, Mark, *op. cit.*

[48] *Ibid.*

[49] *Ibid.*

[50] Xie, John, *op. cit.*

[51] Andrews, Mark, *op. cit.*

[52] Xie, John, *op. cit.*

system that emphasises imaging rather than PNT functionality offers a messaging service).[53] Larry Wortzel, commissioner on the US–China Economic and Security Review Commission and a boat owner, explained about the advantageous feature of the SPOT system:

> In the event of an emergency, I can generate an emergency locator signal that will allow the Coast Guard or marine rescue to come to my aid.… The SMS text system will allow me to send pre-programmed, very short, text messages to a specific number.[54]

Dual Use Technology: BeiDou 3 was built based on commercialization with the basic positioning services free but is projected to trawl back returns of up to US$59 billion yearly as Namrata Goswami, an independent analyst and author specializing in space policy, indicated the following:

> This would include selling its satellite services in an ecosphere were satellite-based communications is critical for an entire economy of a country to work [and that the global navigation market is forecast to grow to $146 billion by 2025].[55]

Blaine Curcio, a space and satcom industry consultant, suggested how Beidou may muster advantages over Galileo/GPS for consumers, not in terms of technical sophistication but in other ways:

> I could imagine two ways: One by subsidizing BeiDou-related equipment, which could allow customers to justify switching from GPS to BeiDou to save money. Two, which is more likely, to market BeiDou as part of a suite of services… For example, if we think about a world 10–15 years down the road, we could imagine China having fleets of autonomous vehicles using the BeiDou satnav standard for location tracking.[56]

---

[53] Andrews, Mark, *op. cit.*
[54] *Ibid.*
[55] *Ibid.*
[56] *Ibid.*

The Chinese state has done its best to increase the use of Beidou in local smartphones systems, fishing boat navigational systems and 6.5 million commercial vehicles, as Namrata Goswami, an independent analyst and author specializing in space policy, pointed out the following:

> BeiDou has brought down costs for farming, logistics, robotics and support to domestic aviation in PNT critical for the local industrial and economic production base [although challenge is whether the Beidou system can provide the services 100 times more accurately with more affordability].... We will have to wait and see if that will be the reality.... But like in other sectors, including 5G, ground solar and rare earth minerals, China could take the lead.[57]

China has launched a broader and lengthier Long March 11 long-range carrier rocket with stronger thrust to put new, heavier payload satellites into space.[58] Beidou is perceived as part of Chinese President Xi Jinping's "Dual Circulation" and "The Infinity Loop" policy with an accent on indigenous technologies, as Heath Sloane, Yenching Scholar at Peking University, explained the following:

> While discourse around 'decoupling' tends to focus on US efforts to reduce its reliance on China, the case of BeiDou shows China's technological disengagement with the USA [perhaps going back to the self-sufficiency guiding principle since 1949.][59]

National Security Issues: There are US experts who believe that it is not possible to decouple dual use technologies into civilian and military applications. Congressional leaders are worried about the potential of Chinese components in some American weapons systems, cyber intrusions into US networks/computers, a scenario laid out in *The Wall Street Journal* article that indicated the Pentagon has "slowly become dependent

---

[57] *Ibid.*

[58] Osborn, Kris, "Lawmakers concerned China may hack, disrupt US military satellite networks" dated 22 July 2020 in Foxnews (downloaded on 22 July 2020). Available at https://www.foxnews.com/tech/lawmakers-china-hack-disrupt-us-military-satellite-networks.

[59] Andrews, Mark, *op. cit.*

upon Chinese industrial output. Asia produces 90% of the world's circuit boards – more than half of them in China."[60]

Specific to GPS technologies, Dr. Larry Wortzel, a commissioner of the US–China Economic and Security Review Commission (USCC) is worried about security issues associated with Beidou and explains his own opinions why the US is also concerned about the mobile phones that tap into the Beidou system:

> All cellular devices, as I understand their function, can be tracked because they continually communicate with towers or satellites.... So just as here in the U.S., there are concerns that police or federal agencies can track people by their cellphones. That can happen. The same is true of a cellphone relying on BeiDou, Glonass and Galileo. The question is: Who are you concerned about being tracked by? [Depending on where the phone is made and what microchips are in the phone] the malware might be embedded in the chips.... That is why the U.S. is concerned about Huawei devices and systems as well as Lenovo computers.[61]

In a Washington conference in March 2020, General James Holmes, Head of United States Air Force (USAF) Air Combat Command, indicated elite American U-2 pilots do not operate/access China's Beidou system as a GPS backup and instead rely on wristwatch systems that receive satellite navigation coordinates from Russian/European systems when GPS is jammed.[62] A 2017 US–China Economic and Security Review Commission (USCC) research study indicated the Beidou system can be a potential portal for cyber attacks:

> BeiDou could pose a security risk by allowing China's government to track users of the system by deploying malware transmitted through either its navigation signal or messaging function (via a satellite communication channel), once the technology is in widespread use [while indicating the experts/ industry professionals has no conclusive knowledge of methodologies in sending malware through a navigation signal].[63]

---

[60] Osborn, Kris, *op. cit.*
[61] Xie, John, *op. cit.*
[62] *Ibid.*
[63] *Ibid.*

For the reasons cited above, the US is utilizing new technologies to protect its communications. The US is launching the Lockheed Martin United Launch Alliance Atlas V rocket from launch complex 41 at the Cape Canaveral Air Force Station with a high-frequency satellite payload on 26 March 2020 at Florida's Cape Canaveral for super-secure communications.[64]

Eventually, power projection and national security will be an important guiding force behind Beidou as Heath Sloane, Yenching Scholar at Peking University, articulated the following:

> Global navigation systems have become the essential condition of modern armed forces.... In addition to its explicit role in enabling accurate navigation, GPS technology allows accurate communications, mapping and targeting, among other applications.[65]

Pakistan (Beidou's first foreign customer) and Thailand are utilizing Beidou for their military applications while Thailand is incorporating the system into its civil service, with other BRI participants as potential customers, as Heath Sloane, Yenching Scholar at Peking University, indicated the following:

> BeiDou is intrinsically linked to 5G and IoT, as they collectively comprise a major component of China's 'Digital Silk Road'.... In addition to the Belt and Road Initiative, the Digital Silk Road is a blueprint to become a global leader in advanced technologies, including artificial intelligence, satellites, smart cities and telecommunications, among others.[66]

China has marketed and incentivized the Beidou to other foreign customers with loans, free services while inking an approximately 2 billion yuan (US$297 million) contract with Thailand in 2013 as the pioneering foreign Beidou customer.[67]

While Thailand and Pakistan are embracing Beidou, others are far more cautious and suspicious. Nowhere is security concerns more acute

---

[64] Osborn, Kris, *op. cit.*

[65] Andrews, Mark, *op. cit.*

[66] *Ibid.*

[67] *Ibid.*

than in Taiwan. China's rival across the straits, Taiwan, which it considers a runaway province and often addresses the island as "Chinese Taipei" or "Taiwan province" has legislated laws in 2016 that take into consideration the potential for the utilization of two-way communication capabilities for cyberattacks and therefore compels the civil service to not use Beidou-compatible smartphones' navigation system.[68] The Taiwanese Ministry of Science and Technology public report indicated Taiwanese individual utilizing mainland-made hand-phones may be revealing information to Chinese agencies through embedded malware and therefore the Ministry has requested national defence agencies track Beidou transmitted signals and spot anomalies within the shortest time possible:

> Because the Chinese BeiDou satellite positioning system has two-way information sending and receiving function and malicious programs could be hidden in the navigation chip of the mobile phone, operating system or apps, the use of BeiDou-enabled smartphones could face security risks.[69]

As an integral component of every aspect of military ops, Beidou's larger scale and size have huge consequences for the tech industry and national security.

There were rare instances of cooperation between China and the US. For example, Beidou received crucial Washington assistance in frequency band availability in 2017 when it was not available because, under the "first come, first served" principle, GPS took up majority of the GPS spectrum as the US was the pioneering country to initiate broadcasting in those frequencies.[70] China therefore needed Washington's approval to utilize this limited resource and, after three years of discussion, both sides agreed in December 2017 for the interoperability of BeiDou civil signals with GPS and the three frequency bands through which BeiDou satellites send navigation signals are located adjacent to or even within GPS frequency bands.[71]

Beidou Chief Designer Yang Changfeng declared on state-owned new channel *CCTV* that China was "moving from being a major nation in

---

[68] Xie, John, *op. cit.*
[69] *Ibid.*
[70] *Ibid.*
[71] *Ibid.*

space to becoming a true space power."[72] The Beidou navigation system achievement is so iconic for the authorities that a BDNSS mock-up was displayed in Hunan Changsha China during China's Space Day 2019 on 23–24 April 2019.[73] Despite rapid progress made by China's Beidou, they are still behind the advanced scientific nations in satellite technology capabilities as Blaine Curcio, a space and satcom industry consultant, explained the following:

> The Chinese space industry can produce most of the same types of satellites that the West can produce, but at a somewhat lower level of technological sophistication.... The most advanced Western-made communications satellites offer hundreds of Gbps [in signal speed]. The most advanced Chinese communications satellites are around 30–50 Gbps. This puts them probably 7–10 years behind where the West is right now in communications satellites, for example.[74]

Therefore, China's advantage lies in cost-effectiveness for global customers, as Heath Sloane, Yenching Scholar at Peking University, pointed out: "If BeiDou can provide a highly accurate and cost-competitive alternative to other GNSS systems, it may present an enticing option irrespective of any shortcomings."[75]

Meanwhile, the US is moving on to new and less vulnerable technologies. For example, US-based private company Ligado is developing a terrestrial 5G type network that may be more resilient, quicker and efficient than GPS in location, navigational, networking and targeting data, perhaps useful as a response to the debate whether there are potential alternatives, replacements or upgrades.[76] An anonymous senior US government official mentioned the following: "If Ligado's system is allowed to broadcast, it breaks up GPS and those devices that use GPS have to be retrofitted.... Ligado uses a 10-watt signal, which is stronger than GPS. It generates too much extra energy and overloads a lot of the GPS receivers and interferes with GPS chips."[77] Ligado's terrestrial wideband wireless

---

[72] *Ibid.*
[73] *Ibid.*
[74] Andrews, Mark, *op. cit.*
[75] *Ibid.*
[76] Osborn, Kris, *op. cit.*
[77] *Ibid.*

transmitters operating at 10 watts are more resilient and closer-in than space-based GPS signals that weaken when beamed back towards Earth and, if the system works, it could have a global customers for the same reasons that the US has, just as BDNSS is constantly launching satellites to be that alternative to GPS.[78]

However, there are US critiques of the Ligado system. The US Department of Defense (DoD) announced the following: "There is no need to put GPS at risk. Mid-band spectrum for 5G exists, and DOD is working with industry on a dynamic spectrum-sharing framework. Ligado's proposal is unnecessary. Ligado's proposed network lacks the bandwidth, power or global ecosystem to deliver robust 5G services. The only beneficiaries are Ligado shareholders."[79] Currently, Ligado's land-based 5G network using Mobile Satellite Service Band is not integrated/operational but it already has the Pentagon elites worried and there is Congressional lobbying not to FCC-approve Ligado's technical initiatives based on national security reasons.[80]

There are also worries that Chinese technologies could break into the global technical communications market, US/NATO/Allied networks as a senior US government official noted the following:

If BeiDou gets into our military systems, it could provide false timing or bad position information.... It would allow nefarious Chinese technology to infiltrate U.S. systems along the supply chain to include command and control systems and computer networks. [While contractors supporting the U.S. market could be vigilant regarding what kinds of technologies or systems are integrated, the subcontracting process is often so detailed, small-scale and elaborate that it could make things difficult, if not impossible to track. In fact, BeiDou elements are already in use in some commercial U.S. technologies.] The FCC has not banned the use of BeiDou systems in things like some cellphones and computer chips. It is not being used in U.S. military equipment, but the threat could easily come up.[81]

---

[78] *Ibid.*
[79] *Ibid.*
[80] *Ibid.*
[81] *Ibid.*

On the other hand, Ligado argues that its system is beneficial to US satellite networks, its military applications and dual-use commercial technology:

> The Ligado SkyTerra 1 satellite provides coverage and capacity that can be dynamically allocated in response to emergency situations or for any additional bandwidth needs. Advanced ground-based beamforming capabilities enable flexible management of network resources, forming hundreds of beams when and where they are needed throughout North America. [Some additional technical reasons why its networks may exceed the capacities of current GPS, potentially inspiring a new competition for military satellite technology.] The Ligado satellite network leverages lower mid-band spectrum, which is not only less susceptible to rain fade and blockage from dense foliage, but can also be tightly integrated with terrestrial networks to deliver seamless satellite and terrestrial coverage.[82]

Impact on Neighbours: China's space technologies have both invoked competitive as well as inspirational instincts among its neighbours. For Japan, China is not necessary a threat in every aspect for Japan's space endeavours. China in fact has been an inspiration to Japanese efforts to push on with space explorations. Hong Kong media *South China Morning Post* reported that Japan's space exploration in fact have been encouraged and motivated by the achievements and success of China's Jade Rabbit lunar rover mission in 2013 which increased Chinese soft power when it exported its space technologies and specialized aerospace engineering training programmes to other countries.[83] This was not the first time foreign achievements have inspired the Japanese space programme. The 1957 Sputnik "shock" as well as the successful US Apollo Programme during the Cold War also inspired the Japanese government and industries to catch up with its peers.[84]

In the security arena and in the area of counterstrike, America's rivals are looking at kinetic and non-kinetic weapons like funding laser systems and satellites that can track US space vehicles, currently in development

---

[82] *Ibid.*

[83] Ryall, Julian, "Amid rivalry with China, Japan is aiming for the moon — and beyond."

[84] Akimoto, Daisuke, "The Evolution of Japan's Space Strategy and the Japan–US Alliance."

in Russia, China, Iran and North Korea.[85] The pull-out from the Anti-Ballistic Missile Treaty in 2002 frees the United States to develop a missile defence system to intercept North Korean missiles, Chinese intermediate-range missiles (with conventional and nuclear warheads) and Russian short-range (500 km) Iskandar ballistic missiles. The missile shield covers allies like South Korea and Japan from missile strikes as well.[86]

Commercially, China's rocket launches are targeted at serving developing economies like Nigeria, Venezuela, Sri Lanka and Pakistan while the United States, the United Kingdom and France have selected India's ISRO rockets to launch their satellites into space[87] though China and India have equal ambitions to capture the global communication satellite market. Currently, India is behind China in the commercial launch market with capabilities to launch 8–12 satellites annually, compared to China's 18–20[88]; however, India is determined to work with the West and Japan to catch up.

Bibliography

Akimoto, Daisuke, "The Evolution of Japan's Space Strategy and the Japan–US Alliance" dated 28 August 2020 in PacNet #49 (downloaded on 31 August 2020). Available at https://mailchi.mp/pacforum/pacnet-49-the-evolution-of-japans-space-strategy-1170346?e=1dea89bef7.

Andrews, Mark, "Navigating the World" dated 15 March 2021 in the Cheung Kong Graduate School of Business (CKGSB) Knowledge (downloaded on 15 March 2021). Available at https://english.ckgsb.edu.cn/knowledges/china-global-navigation-system-beidou/.

Broad, William J. and David E. Sanger, "China Tests Anti-Satellite Weapon, Unnerving U.S." dated 18 January 2007 in New York Times (downloaded 1 January 2020). Available at https://www.nytimes.com/2007/01/18/world/asia/18cnd-china.html.

---

[85] Mehta, Aaron, "America's adversaries keep investing in weapons to take out satellites."

[86] Sato, Yoichiro, "Missile defense in Japan after the Aegis Ashore cancellation."

[87] Kumar, Chethan, "After Asat, Where do India and China Stand" dated 29 March 2019 in the Times of India (downloaded on 29 March 2019). Available at https://timesofindia.indiatimes.com/india/with-asat-india-ups-its-game-in-space-race/articleshow/68612098.cms.

[88] Kumar, Chethan, *op. cit.*

Goh, Deyana, "China to build and launch Cambodia's first satellite" dated 12 January 2018 in SpaceTech Asia (downloaded on 12 January 2018). Available at https://www.spacetechasia.com/china-to-build-and-launch-cambodias-first-satellite/.

Halappanavar, Abhilash, "The final satellite in the BeiDou system completes an undertaking 20 years in the making" dated 26 June 2020 in The Diplomat (downloaded on 26 June 2020). Available at https://thediplomat.com/2020/06/chinas-answer-to-gps-is-now-fully-complete/.

Hanna, Andrew, "Iran's Ambitious Space Program" dated 29 July 2020 in United States Institute of Peace (USIP) website (downloaded on 29 July 2020). Available at https://iranprimer.usip.org/index.php/blog/2020/jun/23/iran%E2%80%99s-ambitious-space-program.

Kumar, Chethan, "After Asat, Where do India and China Stand" dated 29 March 2019 in the Times of India (downloaded on 29 March 2019). Available at https://timesofindia.indiatimes.com/india/with-asat-india-ups-its-game-in-space-race/articleshow/68612098.cms.

Lim, Tai Wei, "Japan's Space Programme; The Final Frontier" dated 1 October 2020 in NUS EAI Background Brief (Singapore: NUS EAI), 2020.

Mehta, Aaron, "America's adversaries keep investing in weapons to take out satellites" dated 30 March 2020 (downloaded on 30 March 2020). Available at https://www.c4isrnet.com/battlefield-tech/space/2020/03/29/countries-keep-investing-in-weapons-to-take-out-satellites/.

New Scientist and AFP, "Timeline: China's spaceflight history" dated 12 October 2005 in New Scientist (downloaded on 1 January 2020). Available at https://www.newscientist.com/article/dn8144-timeline-chinas-spaceflight-history/.

Osborn, Kris, "Lawmakers concerned China may hack, disrupt US military satellite networks" dated 22 July 2020 in Foxnews (downloaded on 22 July 2020). Available at https://www.foxnews.com/tech/lawmakers-china-hack-disrupt-us-military-satellite-networks.

Pearson, James and Jack Kim, "North Korea appeared to use China truck in its first claimed ICBM test" dated 4 July 2017 in Reuters (downloaded on 4 July 2017). Available at https://www.reuters.com/article/us-northkorea-missiles-china-truck-idUSKBN19P1J3.

Ryall, Julian, "Amid rivalry with China, Japan is aiming for the moon — and beyond" dated 1 July 2020 in South China Morning Post (SCMP) (downloaded on 1 July 2020). Available at https://www.scmp.com/week-asia/economics/article/3091367/amid-rivalry-china-japan-aiming-moon-and-beyond.

Sato, Yoichiro, "Missile defense in Japan after the Aegis Ashore cancellation" dated 1 July 2020 in Japan Times (downloaded on 1 July 2020). Available at https://www.japantimes.co.jp/opinion/2020/07/01/commentary/japan-commentary/missile-defense-japan-aegis-ashore-cancellation/.

Stokes, Mark, Gabriel Alvarado, Emily Weinstein and Ian Easton, China's Space and Counterspace Capabilities and Activities (The US–China Economic and Security Review Commission), 2020.

Xie, John, "China's Rival to GPS Navigation Carries Big Risks" dated 8 July 2020 in Voice of American (VOA) (downloaded on 8 July 2020). Available at https://www.voanews.com/a/east-asia-pacific_voa-news-china_chinas-rival-gps-navigation-carries-big-risks/6192460.html.

Xinhua, "China to launch communication satellite for Indonesia" dated 1 April 2020 in Xinhuanet (downloaded on 1 April 2020). Available at http://www.xinhuanet.com/english/2020-04/01/c_138937376.htm.

Xinhua News Agency, "China strengthens international space cooperation" dated 23 April 2018 in Space Daily (downloaded on 23 April 2018). Available at https://www.spacedaily.com/reports/China_strengthens_international_space_cooperation_999.html.

Xinhuanet, "China Focus: China's space tech-based robots find wider application on Earth" dated 22 August 2019 in Xinhuanet (downloaded on 22 August 2019). Available at http://www.xinhuanet.com/english/2019-08/22/c_1383 29594.htm.

Zhao, Lei, "China launches second remote-sensing satellite for Venezuela" dated 10 October 2017 in China Daily (downloaded on 10 October 2017). Available at https://www.chinadaily.com.cn/china/2017-10/10/content_330 58430.htm.

Zhu, Sirui, "China's Long March to Space" dated 27 December 2019 in Reuters (downloaded on 27 December 2019). Available at https://graphics.reuters.com/SPACE-EXPLORATION-MOON/0100B0BH0NZ/index.html.

Printed in the United States
by Baker & Taylor Publisher Services